I0160212

Milo Darwin Burke

Brick for Street Pavements

Milo Darwin Burke

Brick for Street Pavements

ISBN/EAN: 9783337376499

Printed in Europe, USA, Canada, Australia, Japan

Cover: Foto ©berggeist007 / pixelio.de

More available books at **www.hansebooks.com**

Brick for Street Pavements.

AN ACCOUNT OF TESTS MADE OF BRICKS AND PAVING
BLOCKS, WITH A BRIEF DISCUSSION OF STREET
PAVEMENTS AND THE METHOD OF CON-
STRUCTING THEM.

BY

M. D. BURKE, C. E.

CINCINNATI:
ROBERT CLARKE & CO.
1892.

PREFACE.

A large part of the contents of this pamphlet was contained in a report made to the village authorities of tests of material to be used in paving streets in Avondale, where the writer was employed as village engineer. The investigation then made was as thorough as the time and means at hand would justify. Inquiries for the results of the work have been so numerous, and requests for the same have been so frequently repeated, that it has been deemed advisable to publish the work in this form.

If any information or suggestions herein contained shall aid in the construction of better pavements, or prevent the waste of money upon bad ones, or shall bring people to a consideration of placing municipal improvements under systematic control and direction, or shall cause manufacturers to see that their true interest will best be subserved by placing *only* good material upon the market, then this little pamphlet will have served a useful purpose.

M. D. BURKE.

CINCINNATI, *March* 16, 1892.

.

BRICK FOR STREET PAVEMENTS.

THE SAMPLES SUBMITTED FOR TESTING.

It having been decided that the Main Avenue pavement should be constructed of some form of clay product, a series of tests of the various materials in market was deemed advisable in order to aid in selecting the proper variety, and, if found practicable, fix a standard by which the different varieties might be adjudged and accepted or rejected, as their qualities and powers of resistance would determine. Accordingly a circular letter was addressed to manufacturers and dealers, requesting them to ship to my office, at 41 Pike Building, Cincinnati, Ohio, samples consisting of twenty or more of each of the varieties of the bricks or blocks manufactured or sold by them for street paving purposes, to be tested. In response to the circular letter, samples were kindly furnished by the following parties, and numbered as below:

1. Lithonia Georgia Granite, John Regan, contractor, city.

2. West Virginia Brick Co., Charleston, W. Va., H. C. Bruce, President.

3. The Diamond Brick and Terra Cotta Co., Kansas City, Mo.

4. The Pittsburg Sewer Pipe and Fire Clay Co., New Brighton, Pa.

5. Canton Brick Co. (red granite street pavers), Canton, Ohio.

6. The Royal Brick Co. (iron rock pavers), Canton, O.

7. Purington Paving Brick Co., Galesburg, Ill.

8. The United States Fire Clay Co., New Lisbon, Ohio ; M. R. Coney, agent.

9. The Middleport Granite Brick Co. (Hallwood Block), Middleport, Ohio.

10. L. B. Townsend & Co. (Townsend Paver), Zanesville, Onio.

11. The Brazil Paving Block, Brazil, Ind., L. H. McCammon Bros., agents.

12. The Jones Paving Block, Zanesville, Ohio, L. H. McCammon Bros., agents.

13. The Ohio Paving Co. (Hallwood Block), Columbus, Ohio.

14. The John Porter Co., New Cumberland, W. Va.

15. The New York Paving Brick Co., Syracuse, N. Y.

16. Hallwood Block Granite Brick, manufactured by Tennessee Paving Brick Co., Robbins, Tenn.

In this report the varieties are represented by the numbers above given, and the separate specimens by letters. Of each variety, except Nos. 1 and 16, ten bricks or blocks were used in making the various tests. A number always indicates the same variety, but only when the number and letter are the same, does it mean the same brick. Thus No. 1 always means a granite block, yet No. 1 A and No. 1 E are different blocks, but both Georgia granite.

It was deemed advisable to ascertain, first, the essential chemical ingredients ; second, the ratio of absorption ; third, crushing strength ; fourth, transverse strength ; fifth, the resistance to abrasion and impact. The Riehle testing machine of Messrs. Otten & Westenhoff, appearing to be the most readily available, these gentlemen were employed, not only to make the crushing and transverse tests, but also to make the chemical analysis and determine the specific gravity and ratio of absorption of the cubes which were prepared for the crushing test.

HOW THE SPECIMENS WERE PREPARED.

The testing of a single specimen of any kind being deemed insufficient, it was determined that three cubes of each variety be prepared for ascertaining the crushing strength. Accordingly three bricks or blocks of each variety were taken at random, and sent to the marble works of Joseph Foster & Sons, placed in the mill and sawed in two lengthwise, the cut being made so as to leave one piece about $2\frac{1}{2}$ inches in width. This piece was then cut by the saws transversely, so as to approximate as nearly as practicable to a 2-inch cube from the interior of the brick. The roughly sawn cubes were then placed upon a rubbing bed and worn down to the required size, with parallel and equal faces. Three such cubes were made of each variety. Of the granite, A and B were made from one block; C from another. Number 16, the three cubes were made from a single block, as of this number but three blocks were furnished. In making No. 15, which is of a very refractory material, the saw was unfortunately deflected in such a manner that cubes could not be obtained from two of the pieces, B and C, and they were rubbed down to two inches square by $1\frac{1}{2}$ inches high.

Subsequent developments show that it would have been advisable to have made $1\frac{1}{2}$ inch cubes instead of 2 inch. The testing machine used has a capacity of 52,000 pounds. It was thought that very few, if any, of the specimens would show an ultimate crushing strength exceeding 13,000 pounds per square inch, but in this we were mistaken.

By the same process above described, four "granite

bricks" were manufactured; that is, four pieces of granite
2¼x4x8 inches were made, to be subjected to the same tests
as the bricks, in order to compare the resistance of the clay
products with a standard paving material. It will be observed that great care was taken in this work in order to
preserve the material of each specimen intact and to prevent injury to it in any way. No cutting with chisels or
spalling was permitted. Each specimen was numbered and
lettered and its identity preserved throughout the entire
series of tests. The cubes were used for three purposes:
First, for obtaining the ratio of absorption; second, specific
gravity, and third, the crushing strength of the material.
Another absorption test was made with whole bricks, and
in some instances the percentages obtained differed materially. There are two reasons for these differences: A single
cube only was used and it was immersed but twelve hours.
For some of this material probably this length of time was
too short for it to absorb all the water that it would ulti-
mately take up, but generally the percentages obtained by
immersing the cubes was materially higher than that obtained from the whole bricks, which is a result to be expected when it is remembered that the outer portions of
the bricks were in several cases salt-glazed and were generally more dense and burned harder than the interior portion
from which the cubes were obtained. The specimens subjected to crushing were lettered A, B, C, those lettered A
being used for the absorption as well as the crushing test.

    No essential preparation of the specimens for the re-
maining tests was necessary, They were all kept in a steam-
heated room from the time of their arrival until used, which
was about one week for the latest arrivals, and about four
weeks for the earliest. All would be classed as perfectly
dry. All adhering sand or dust was carefully brushed from

those tested for absorption or abrasion before weighing them. In selecting those used for ascertaining the transverse strength, perfect specimens, showing no fire cracks or other defects, were taken. In fact the manufacturers or agents had generally done the selecting and packing with such care that no outside defects were visible, except as noted for No. 10.

DESCRIPTION OF SPECIMENS SUBMITTED.

No. 1. Application was made to the Southern Granite Company for specimens of their material, but none was received, accordingly the samples used were obtained from Mr. John Regan, contractor, who was using Lithonia granite in paving a street, and the cubes and specimen bricks were sawed from the interior of the blocks, rubbed down, without the use of hammer or chisel, to the dimensions given as above noted, in the preparation of the specimens.

No. 2 is a hard burned brick manufactured from plastic clay and burned in the ordinary clamp kiln. It is about $2\frac{1}{3}''x3\frac{3}{4}''x8''$, and if closely laid, about sixty-five of them would pave a square yard. Its record can be traced through the various tests in the following tables by its number. Its history as a street paver is said to be quite satisfactory in some localities, but it should not be used where the traffic is very considerable.

No. 3 is manufactured from a shale or plastic clay which readily vitrifies. In size it averages $2\frac{1}{4}''x3\frac{5}{8}''x8''$, and about sixty-eight will be required for each square yard of pavement. The chemical analysis, as given in table No. 1, does not show that it contains an objectionable amount of lime, but other tests unmistakably manifest its presence in form and quantity to a highly detrimental degree. The brick is very hard and dense, ranking third in specific gravity, but it is rather small and quite brittle, the fracture being conchoidal, and it will yield to the trituration of animals' shoes rather than the grinding of the wheels of vehicles. I have no knowledge of its record in actual service as a street paver.

No. 4 is what is known as a fire-clay brick. In color it is an orange buff. The average size is $2\frac{1}{2}''$x$4\frac{1}{8}''$x$8\frac{1}{2}''$. About fifty-nine would lay a square yard of pavement. It absorbs water rather freely, but not to a greater extent than many acceptable building stones, and in all the other tests its standing is good.

Nos. 5 and 6 are practically the same brick, manufactured from a shale or indurated clay. In color they are a dark red. They average $2\frac{1}{2}''$x$4\frac{1}{8}''$x$8\frac{3}{8}''$, and about fifty-nine of them will pave a square yard. The samples tested are all burned quite hard, but not in all cases to vitrification; hence while they show great transverse strength, and resist abrasion well, they are comparatively low in crushing strength and there is great variation in the percentage of absorption. Their record in actual use is quite satisfactory, but I have no statistics of the severity of the traffic to which they have been subjected.

No. 7 is of a dark red color, $2\frac{1}{4}''$x$3\frac{5}{8}''$x$7\frac{3}{4}''$ in size. About sixty-nine will be required for each square yard of pavement. In all the tests this ranks among the best of the red bricks, and its record under moderate traffic is good. An increase in size so as to afford a greater weight of pavement would appear to be prudent if it is to be used under heavy traffic.

No. 8 is a fire-clay brick of a light buff color, $2\frac{9}{16}''$x$3\frac{15}{16}''$x$8\frac{7}{16}''$, of which about fifty-five will pave a square yard. The material of which it is composed has not been very finely ground nor very thoroughly compressed. As a consequence it has a low specific gravity, a moderately high rate

of absorption, and is outranked by several other varieties in
the other tests, but in the uniform quality of each individual
brick as compared with the others of its kind, it stands at
the head of the list.

No. 9. In chemical constituents this coincides more
nearly with No. 2 than any of the other samples submitted,
yet the treatment of the material has been so different that
the results are in no respect similar. It is a glazed Hall-
wood block, $2\frac{9}{16}''x3\frac{7}{8}''x8\frac{1}{2}''$, and about fifty-five of them will
lay a square yard of pavement. The glaze is said to be a
natural, and not a salt glaze. The corners are rounded to
about one-half inch radius, and two $\frac{1}{4}''$ groves extend lengthwise around the block near its middle. The clay has been
finely ground and completely vitrified, but inasmuch as it
contracts greatly in burning, the blocks are liable to show
cracks on the outside or cavities on the inside. When
broken the blocks show an even dense texture of a dark
brown color, and, were the defect above noted remedied
(which it would appear to the writer, could be readily done),
they would be greatly improved for street paving purposes.

No. 10 had been assorted to some extent when they came
into my hands, as more than one-fourth of the bricks had
been broken in the box. The average size is $2\frac{5}{12}''x4''x8\frac{7}{12}''$,
and fifty-eight of them will lay a square yard of pavement.
They are dark brown in color, with corners rounded to
about one-fourth of an inch radius; burned exceedingly
hard, although they do not present the melted appearance
of most of the vitrified bricks. While this brick has great
hardness, with sufficient density for all practical purposes,
and even a high degree of tenacity under a steadily applied
stress, yet it possesses great brittleness, and when subjected
to shocks, shows a tendency to spall badly. Could the quality of toughness be given it without materially diminishing

its hardness, no essential of a desirable brick paver would be lacking.

No. 11 is manufactured from a clay found in the coal measures but not termed a fire-clay. It is a salt-glazed block $2\frac{3}{4}''$x4''x$9\frac{1}{8}''$ in size, and each square yard of pavement will require about forty-eight blocks. The corners are slightly rounded. Little fault can be found with the chemical ingredients, although an additional amount of iron would be in better proportion to the quantity of alkalies present. The form and size of block has been well chosen if such a thickness can be properly burned, but the mechanical work of preparing the material and forming the block has been indifferently done and the burning decidedly *underdone*.

No. 12, is of well chosen dimensions, being $2\frac{5}{8}''$x$4\frac{1}{8}''$x9'', and fifty-one blocks will lay a square yard of pavement. The material of which it is composed is about the same as that of which No. 10 is made, namely, a mixture of shale and clays found in the coal measures. The corners are rounded to about one-fourth of an inch radius. The blocks are repressed, with five grooves passing transversely nearly across one side, and eight diagonally nearly across the other. The sand, or possibly the oil, used in repressing, serves to give it a reddish brown color. Giving the block a form such that it should always be placed with the same side up appears to be a refinement hardly required in practice. The material might have been made into a good paving block, but it was not. The clays were not ground sufficiently fine, neither was the burning carried to such an extent as to produce a block that would withstand the abrasion of street traffic. The samples tested were obtained from an agent, and not directly from the manufacturers.

No. 13, is salt-glazed, corners rounded to about one-half

inch radius, with two grooves about one and one-half inches apart, passing lengthwise around it near the middle. About forty-six blocks will pave one square yard. It shows a higher percentage of iron than any other specimen analyzed, but appears to be mainly composed of a plastic clay, possibly indurated, which has been finely ground and skillfully combined. In the process of burning or vitrification, the iron and alkalies have combined so as to render the material practically impervious to moisture, but it has not quite as high a specific gravity nor the strength that should be obtained with this material. Its record in practical use is very good, and under any thing like fair treatment will give satisfactory results.

No. 14, is a repressed fire-clay brick, with corners rounded to about three-sixteenths of an inch radius. The average size is 2⅓″x4⅓″x8½″, and sixty of them will pave a square yard. The material is very similar to No. 4, but the repressing has given it advantages in some particulars. It is of a buff color. It has been used as extensively for street paving as any other variety tested, and under moderate traffic its record has been satisfactory. A result appears to be attainable with this material which is not always secured, but when it is, leaves but slight room for apparent improvement in the manufacture of paving blocks. It is obtained by the fusion of the iron with the silica when acted upon by the alkalies or other fluxes, in the process of burning or vitrification, producing a block, which, when broken, presents a gray metallic or granitic texture, showing no traces of cleavage or granular structure, and perfectly uniform throughout. Of the sample bricks of this variety tested, about seventy-five per cent were completely vitrified as here described, while the others presented a granular appearance, either throughout or in the central portion, but they were all very hard burned. The cube used for obtaining specific gravity and percentage of absorption was but partially fused, hence it shows great affinity for moisture. This fusion does not appear to add materially to the strength, but it does lessen the amount of absorption without materially increasing the brittleness.

No. 15, is drab or brown in color, $2\frac{1}{8}''x3\frac{9}{16}''x7\frac{5}{8}''$ in size, requiring about seventy-five of them to lay a square yard of pavement. The clay from which this brick is made is evidently largely formed from the disintegration of limestone rocks. All the samples were thoroughly vitrified, and the product is an exceedingly refractory substance. Could the lime be eliminated from it before burning, the bricks would be as nearly indestructible as could be desired, but every brick tested manifested the presence of lime in quantity sufficient to impair its durability.

No. 16, is made from a shale or clay found in the coal measures, although not described as a fire clay. But three of these blocks were obtained for making the tests, and each was quite a perfect specimen of its kind. This is a Hallwood block, corresponding in dimensions with No. 13, and if the samples fairly represent the product of the kilns, pavements properly laid with this material will be both durable and satisfactory.

How the Tests Were Made.

Absorption.—Of the cubes prepared as before described, the one lettered "A" of each variety was placed in a drying oven and maintained at a temperature of 212° Fahrenheit for twelve hours, in order to drive off any contained moisture. Each was then accurately weighed. The figures obtained are found in the second column from the left of Table No. 2. It was then immersed in water, where it remained for the succeeding twelve hours, when it was taken out, the adhering moisture wiped off, and again weighed, the results being noted in the third column from the left in the same table. At this time it was also weighed in water, these last weights being found in the sixth column of Table No. 2.

For a further test of absorption two whole bricks of each variety (except as noted in the table) were selected at random, lettered G and H respectively, placed on an ordinary counter scale weighing to quarter ounces, the results noted, and are found in the ninth column from the left in table No. 2. They were then placed in water and kept entirely submerged for seventy hours, when they were taken out, dried with a towel, again weighed, with results found in the tenth column of the same table. These results are only as accurate as the scales were, but the test can be readily repeated at any time, and will give a fair practical measure of the absorption to be expected from whole bricks in a similar length of time. The samples had been kept in a steam-heated room during the preceding ten days, and the dust and sand were carefully brushed from them before the first weighing.

Crushing Strength.—For determining the crushing strength the cubes were carefully measured, their upper and lower surfaces protected by a thickness of blotting paper, and they were subjected to pressure in a Riehle Testing Machine of 52,000 pounds capacity. The results obtained will be found in table No. 3. No visible effect was produced upon the granite except that " B " was very slightly spalled, as noted. No. 15 " C," which was about one and one-half by two inches, was set on edge after being tried the other way, and a pressure exerted exceeding 17,000 pounds per

square inch, but it could not be crushed, although it spalled
at one corner at a little over 7,000 pounds per square inch.

Transverse Strength.—The transverse strength was as-
certained in the same machine in the following manner:
Three bricks of each variety (except as noted) were chosen
and lettered D, E and F. The lower knife edges were ad-
justed at precisely six inches apart, the upper knife edge
being placed centrally between them. Each specimen was
carefully measured and its surface protected from direct
abrasion at the points of bearing by two or three thicknesses of blotting-paper, and the weight required to break
it carefully noted. These results are found in table No. 4.

ABRASION AND IMPACT.

The method adopted for determining the relative resistance to abrasion and impact was that which is commonly known as the rattling test. A cylinder about six
feet in length by twenty-eight inches in diameter, contain-
ing pieces of cast iron, varying in weight from one to six
or eight pounds, and in the condition in which it is ordinarily used for cleaning castings, was selected for this purpose.
Four bricks of each variety (with the exceptions noted in the
table) were selected and lettered G, H, I and J, two of the
specimens, G and H, having been soaked for the preceding
seventy hours. The specimens were carefully weighed and
all placed in the rattler at one time. Billets of wood were
put in with them, as is ordinarily done in cleaning castings, to prevent breaking. The cylinder was revolved at a
speed of about twenty-four revolutions per minute, and at
the end of one thousand revolutions was stopped, the specimens taken out, weighed and the loss of each noted. The
object of making this first test in this particular manner
was to wear away the sharp angles or corners and bring

each piece as nearly as practicable on a level footing with
its fellow for that which was to follow. This first test,
therefore, was intended more to equalize the several pieces
than to measure their actual wear.

On the following day the specimens were again placed
in the cleaner, omitting the protecting billets of wood. The
object now being to observe the survival of the fittest.
The cylinder was given three thousand revolutions, occupy-
ing something over two hours in time, and although all the
pieces were in at the same time, excepting a fragment of No.
15 " I," which had broken off in the former test and was in-
advertently omitted; there was ample room for motion and
the loss in weight of every piece was a measurable quantity.
The individuality of some specimens was lost, as the marks
were in some cases almost entirely worn away, but there was
no trouble in identifying the varieties; hence, in tabulating
the results of this work the percentage of loss in the second rattler test is given for each variety. This will be found
in detail in table No. 5.

With the completion of the second rattler test closed
the actual work upon the specimens, and the labor of col-
lating the information obtained was commenced. Further
investigation would appear to be desirable, especially some
test that will more nearly resemble the attrition of the shoes
of animals in passing over the pavement than do any of the
tests that were made. But the information now gained
would appear to justify seeking that knowledge by a practical test of the brick in the street pavement itself.

2

TABLE No. 1.—CHEMICAL ANALYSES.

No. of Specimen.	Silica.	Alumina.	Peroxide of Iron.	Lime.	Magnesia.	Alkalies.	Undetermined.	Total Percentages.
2	73.32	14.82	8.34	0.70	0.99	2.26	....	100.43
3	64.37	19.73	9.07	0.82	2.32	1.89	1.80	100.00
4	67.36	22.05	5.61	0.86	0.36	2.70	1.06	100.00
5	67.65	18.36	8.34	0.80	1.02	2.58	1.25	100.00
6	68.12	18.63	8.53	0.68	0.71	2.58	0.75	100.00
7	68.69	17.95	7.25	0.76	1.47	2.83	1.05	100.00
8	64.08	25.32	5.44	0.30	0.29	0.63	3.94	100.00
9	71.57	17.06	8.34	0.50	0.58	0.56	1.39	100.00
10	61.80	20.76	8.70	1.38	1.09	1.44	4.83	100.00
11	77.67	14.77	3.63	0.38	0.27	2.43	0.85	100.00
12	65.08	22.39	7.97	0.62	0.74	2.33	0.87	100.00
13	66.30	18.62	9.78	0.40	0.84	1.89	2.17	100.00
14	69.02	22.07	4.53	1.70	0.38	1.34	0.96	100.00
15	67.67	11.67	6.53	12.74	0.95	0.80	....	100.36
16	70.57	15.19	7.97	0.78	0.32	1.15	4.02	100.00

Undertermined is water manganese oxide and possibly some titanic acid.

OTTEN & WESTENHOFF,

Chemists.

TABLE No. 2—ABSORPTION AND SPECIFIC GRAVITIES.

Single Tests of 2″ Cubes 12 Hours in Water. Weights in Grammes.

No.	Weight before immersion.	Weight after immersion.	Weight of water absorbed.	Percentage of absorption.	Weight in water.	Specific gravity.
1	344.50	345.60	1 10	0 32	213.75	2 61
2	272.40	284.70	12.30	4 52	159.39	2 17
3	298.87	302.80	3.93	1 31	175.20	2 34
4	280.84	289.00	8.16	2.91	160.20	2 18
5	287.40	290.00	2.60	0.90	164.90	2 29
6	260.28	270.20	9.92	3.81	150.60	2 18
7	298.80	302.20	3.34	1.12	172.60	2 31
8	263.30	269.50	6.20	2 35	142.15	2 07
9	287.15	291.40	4.25	1 48	163.70	2 25
10	287.70	289.60	1.90	0 66	164.10	2 29
11	278.80	283.60	4.80	1.72	157.68	2 21
12	287.70	293.00	5.30	1 84	166 48	2.27
13	262.35	263.10	0.75	0 29	139.40	2 12
14	269.25	277.30	8.05	2 99	151 00	2 13
15	293.20	294.90	1.70	0.58	167 10	2 29
16	300.45	305.20	4 75	1 58	179 60	2 39

Whole Bricks 70 Hours in Water. Weights in Ounces.

Letter.	Weight before immersion.	Weight after immersion.	Weight of water absorbed.	Percentage of absorption.	Mean of the two bricks.
E	116 75	116 75	0 0	0 0	Only one specimen.
G	89 00	94 25	5 25	5.90	6 88
H	89 00	96 00	7 00	7.86	
G	91 50	92 25	0 73	0.82	0 82
H	90 75	91 50	0 73	0.82	
G	113 75	118 00	4 25	3.74	2 765
H	111 73	113 75	2 00	1.79	
G	112 73	114 00	1 25	1.11	1 88
H	113 00	116 00	3 00	2.65	
G	109 25	109 75	0 30	0.46	0 34
H	113 75	114 00	0 25	0.22	
G	87 25	87 75	0 50	0.57	0.285
H	88 50	88 50	0 0	0 0	
G	102 25	103 50	1 25	1.22	1 345
H	101 75	103 25	1 50	1.47	
G	106 75	108 50	1 75	1.61	1 655
H	105 00	106 75	1 75	1.67	
G	108 75	109 00	0 25	0.23	0 23
H	110 00	110 25	0 25	0.23	
G	124 50	126 00	1 50	1.20	1 395
H	111 73	114 00	2 25	1.50	
G	128 50	129 50	1 00	0.78	0 585
H	127 50	127 50	0 50	0.39	
G	133 75	133 75	0 0	0.19	0 095
H	133 50	133 75	0 25	0 0	
G	106 75	107 73	1 00	0.93	0 935
H	106 25	107 25	1 00	0.94	
G	76 50	76 50	0 0	0 0	0 0
H	76 25	76 25	0 0	0 0	
H	120 25	121 73	1 50	1 24	Only one specimen.

TABLE No. 3.—Crushing Strength.

SPECIMEN NO.	LETTER	SIZE IN INC. Height	SIZE IN INC. Length	SIZE IN INC. Br'dth	AREA IN SQUARE IN.	SPALLED AT IN POUNDS.	SPALLED AT IN POUNDS PER SQ. INCH.	CRUSHED AT IN POUNDS.	CRUSHED AT IN POUNDS PER SQ. INCH.	REMARKS.
1	B	2.	2.	2.	4.	40480	10120		....	Did not crush at 52000 lbs.=13000 lbs. per □".
2	A	1.97	1.07	1.94	3.82	29600	7749	35500	9293	Soaked 12 hours.
2	B	1.94	1.94	1.94	3.76	28000	7447	35920	9553	
2	C	1.94	1.97	1.94	3.82	....	....	..		Did not spall at 52000 lbs. = 13613 lbs. per □".
3	A	1.97	1.97	1.97	3.88	32290	8322	37720	9722	Soaked 12 hours.
3	B	1.94	1.97	1.94	3.82	15200	3979	48000	12565	
3	C	1.97	1.94	1.94	3.76	18100	4813	32720	8702	
4	A	1.97	1.97	1.97	3.88	33110	8533	43120	11113	Soaked 12 hours.
4	B	1.97	1.97	1.97	3.88	36590	9427	....	....	Did not crush at 52000 lbs. = 13402 lbs. per □".
4	C	1.97	1.97	1.97	3.88	....	....	....	....	Did not spall at 52000 lbs. = 13402 lbs. per □".
5	A	1.94	1.97	1.94	3.82	50000	13089	...	....	Soaked 12 hours. Did not crush at 52000 lbs. = 13613 lbs. per □".
5	B	1.97	1.97	1.97	3.88	38850	10013	39150	10090	
5	C	1.97	1.97	1.97	3.88	28000	7216	38850	10013	
6	A	1.94	1.91	1.91	3.65	19650	5383	23100	6329	Soaked 12 hours.
6	B	1.94	1.94	1.91	3.70	37220	10059	51500	13919	
6	C	1.97	2.00	1.97	3.94	....	....	..	...	Did not spall at 52000 lbs. = 13198 lbs. per □".
7	A	2.00	1.97	1.97	3.88	25120	6471	51600	13300	Soaked 12 hours.
7	B	1.97	1.97	1.97	3.88	....	....	....	....	Did not spall at 52000 lbs. = 13402 lbs. per □".
7	C	1.97	1.97	1.97	3.88	38550	9935	52000	13402	
8	A	1.97	1.97	1.94	3.82	34000	8900	46460	12162	Soaked 12 hours.
8	B	1.97	1.97	1.94	3.82	25220	6602	44650	11688	
8	C	1.97	1.97	1.97	3.88	25030	6451	39570	10198	
9	A	1.97	1.97	1.94	3.82	24000	6283	39820	10424	Soaked 12 hours.
9	B	1.97	2.12	1.97	4.18	26000	6220	48270	11548	
9	C	2.00	1.97	1.94	3.82	17500	4581	39820	10424	
10	A	1.97	1.97	1.97	3.88	50750	13080	....	....	Soaked 12 hours. Did not crush at 52000 lbs. = 13402 lbs. per □".
10	B	1.97	1.97	1.97	3.88	35500	9149	51460	13263	
10	C	1.97	1.97	1.94	3.82	32000	8377	50050	13102	
11	A	1.97	1.97	1.97	3.88	27900	7191	50300	12964	Soaked 12 hours.
11	B	1.97	1.97	1.94	3.82	24000	6283	39400	10314	
11	C	1.97	1.97	1.97	3.88	23880	6155	27250	7023	
12	A	1.97	1.94	1.94	3.76	19700	5239	37330	9928	Soaked 12 hours.
12	B	1.97	1.97	1.97	3.88	14240	3670	17620	4541	
12	C	2.00	1.97	1.94	3.82	10960	2869	28110	7359	
13	A	1.91	1.97	1.97	3.88	28000	7216	46000	12371	Soaked 12 hours.
13	B	1.97	1.97	1.97	3.88	40000	10309	46600	12010	
13	C	1.94	1.97	1.91	3.76	19500	5186	38380	10207	
14	A	1.97	1.97	1.97	3.88	38800	10000	50770	13085	Soaked 12 hours.
14	B	1.97	1.97	1.97	3.88	....	....	...	...	Did not spall at 52000 lbs. = 13402 lbs. per □".
14	C	1.97	1.97	1.94	3.82	....		....	....	Did not spall at 52000 lbs. = 13613 lbs. per □".
15	A	1.97	1.97	1.97	3.88	27770	7157	...	....	Soaked 12 hours. Did not crush at 52000 lbs. = 13402 lbs per □".
15	B	1.50	1.97	1.97	3.88	....	.			Did not spall at 52000 lbs. = 13402 lbs. per □".
15	C	1.50	2.03	1.94	3.94	....				Did not spall at 52000 lbs. = 13198 lbs. per □".
15	C	1.94	2.03	1.50	3.04	35400	11644			Did not crush at 52000 lbs. = 17105 lbs. per □".
16	A	2.	2.	2.	4.00	51660	12915			Soaked 12 hours. Did not crush at 52000 lbs. = 13000 lbs. per □".
16	B	2.	1.97	1.97	3.88	....	....			Did not spall at 52000 lbs. = 13402 lbs. per □".
16	C	2.	2.	2.	4.00	...	....			Did not spall at 52000 lbs. = 13000 lbs. per □".

TABLE No. 4.—TRANSVERSE STRENGTH.

NUMBER.	LETTER.	SIZE IN INCHES. Breadth.	Depth.	Length.	SECTIONAL AREA IN SQ INCHES.	BREAKING WEIGHT IN POUNDS.	MODULUS OF RUPTURE.	AVERAGE OF THREE TESTS.	REMARKS.
							R	R	
1	..	2 25	4.19	6.0	9 43	6590	1501	....	Only one specimen broken. Broke at one of the lower knife edges.
2	D	2.31	3 82	6.0	8 82	4580	1222		
2	E	2.31	3.75	6.0	8 66	6500	1801	}1444	
2	F	2.31	3.875	6.0	8 95	5050	1310		
3	D	2.125	3 625	6.0	7.70	5620	1811		
3	E	2.19	3.625	6.0	7.94	7600	2377	}2010	
3	F	2.125	3.44	6.0	7.31	5400	1932		
4	D	2.50	4.125	6 0	10 31	11680	2472		
4	E	2.50	4 125	6 0	10.31	8000	1693	}2197	
4	F	2.50	4.06	6.0	10 15	11110	2427		
5	D	2 50	3.94	6.0	9 85	13320	3089		
5	E	2.50	4.00	6.0	10 00	11160	2511	}2963	
5	F	2 44	4.125	6.0	10 06	15170	3288		
6	D	2.56	4 06	6 0	10 39	12460	2657		
6	E	2 625	4.00	6 0	10 50	10870	2329	}2494	
6	F	2.625	4 06	6.0	10 66	12000	2496		
7	D	2 25	3 50	6 0	7 875	7020	2292		
7	E	2 31	3 625	6.0	8.37	10250	3525	}2822	
7	F	2.25	3 44	6 0	7.74	7840	2650		
8	D	2 50	3 94	6 0	9.85	8320	1959		
8	E	2.50	3 94	6 0	9.85	7850	1820	}1863	
8	F	2.56	3 94	6.0	10.08	8000	1811		
9	D	2.50	3.875	6 0	9.69	8730	2093		
9	E	2.50	3.875	6.0	9.69	7410	1776	}1672	Round corners, grooved longitudinally.
9	F	2.50	3 875	6.0	9.69	4790	1148		
10	D	2 375	3.91	6 0	9.36	9690	2365		
10	E	2 31	4.00	6.0	9.24	8000	1948	}2299	Round corners.
10	F	2 375	4 125	6.0	9.79	11610	2584		
11	D	2.75	4.06	6.0	11.17	6780	1346		
11	E	2.81	4 00	6 0	11.24	6000	1197	}1195	Round corners.
11	F	2.75	4.00	6.0	11.00	5100	1043		
12	D	2.62	4 06	6.0	10.64	8000	1668		
12	E	2 55	4 19	6.0	10 68	9010	1811	}1666	Round corners, grooved transversely.
12	F	2.55	4 25	6.0	10.83	7770	1518		
13	D	3.00	3.875	6.0	11.62	9760	1850		
13	E	3.06	3.94	6.0	12.06	7830	1483	}1688	Round corners, grooved longitudinally.
13	F	3.00	3.75	6.0	11.25	7640	1630		
14	D	2.44	4.125	6.0	10.06	11660	2527		
14	E	2.375	4 25	6.0	10.09	12710	2667	}2428	Round corners.
14	F	2.375	4.18	6.0	9 93	9640	2091		
15	D	2.06	3.56	6.0	7.33	5750	1982		
15	E	2.06	3.50	6.0	7 21	8000	2853	}2269	
15	F	2.06	3.50	6 0	7.21	5530	1972		
16	..	2.75	3.75	6.0	10.31	7150	1664	....	Only one specimen. Round corners, grooved longitudinally.

For numbers 9, 10, 11, 12, 13, 14, and 16, the dimensions are those of the estimated equivalent rectangular sections.

TABLE No. 5.—ABRASION AND IMPACT.

NUMBER	LETTER	ORIGINAL WEIGHT IN OUNCES.	WEIGHT AT END OF 1ST TEST	LOSS IN 1ST TEST	FINAL WEIGHT.	LOSS IN 2D TEST.	TOTAL LOSS IN OUNCES.	PER CENT LOST IN 1ST TEST	PER CENT LOST IN 2D TEST.	TOTAL PER CENT LOST.	REMARKS.
1	E	116.75	116.0	0.75	110.75	5.25	6	0.6			Soaked 70 hours.
1	F	115.0	114.75	0.25	109.75	5.0	5.25	0.1	4.76	5.17	Soaked 70 hours.
1	G	116.0	115.50	0.50	109.25	6.25	6.75	0.4			"
2	H	94.25	90.25	4.0	63.50	26.75	30.75	4.2			Soaked 70 hours.
2	I	90.0	90.25	5.75	60.50	29.75	33.50	6.0	23.93	26.59	"
2	J	96.75	86.50	1.25	71.0	15.50	16.75	1.4			"
3	G	87.75	86.00	1.75	73.50	12.50	14.25	2.2			Soaked 70 hours.
3	H	92.25	90.25	2.0	67.25	23.0	25.0	1.4	24.14	25.37	"
3	I	91.50	90.25	1.27	80.75	9.50	10.75	2.0			"
3	J	86.0	85.25	0.75	79.50	3.73	5.50	0.9			"
4	G	118.0	116.25	1.75	37.25	48.0	48.75	1.5			Broke and fragments lost.
4	H	113.0	113.25	0.50	103.0	13.25	15.0	0.4	9.42	10.27	Soaked 70 hours.
4	I	113.75	113.50	0.25	104.0	9.25	9.75	0.1			"
4	J	112.0	110.75	1.73	103.25	10.25	11.25	1.5			"
5	G	114.0	114.0	0.75	100.25	10.0	13.75	0.9			Soaked 70 hours.
5	H	116.0	115.0	1.75	103.0	12.25	12.75	1.5	10.05	11.24	"
5	I	114.0	112.25	1.50	101.0	11.75	13.0	1.3			"
5	J	112.0	110.50	1.25	100.25	11.25	11.75	1.1			"
6	G	109.75	108.50	0.75	96.0	12.50	11.50	0.9			Soaked 70 hours.
6	H	111.0	113.0	1.50	102.50	9.30	10.73	1.3	9.86	10.89	"
6	I	112.25	111.25	0.50	101.25	11.25	12.50	0.6			"
6	J	87.75	110.75	1.30	99.50	10.50	10.75	1.4			"
7	G	88.50	87.25	0.50	77.25	12.0	13.25	1.7			Soaked 70 hours.
7	H	87.0	87.75	1.25	71.75	16.0	9.25	1.2	13.01	14.12	"
7	I	88.25	87.0	1.25	73.0	16.75	17.25	1.2			"
7	J	103.50	86.25	1.25	78.50	18.0	17.0	1.2			"
8	G	103.25	102.25	1.25	86.25	16.0	18.0	1.2			Soaked 70 hours.
8	H	101.50	102.0	1.25	85.25	16.75	19.25	1.2	16.31	17.28	"
8	I	100.25	100.25	1.25	82.25	18.0	16.50	0.9			"
8	J	102.50	101.50	1.0	86.0	13.50					"

TABLE No. 5.—ABRASION AND IMPACT.—Continued.

NUMBER	LETTER	ORIGINAL WEIGHT IN OUNCES	WEIGHT AT END OF 1ST TEST	LOSS IN 1ST TEST	FINAL WEIGHT	LOSS IN 2D TEST	TOTAL LOSS IN OUNCES	PER CENT LOST IN 1ST TEST	PER CENT LOST IN 2D TEST	TOTAL PER CENT LOST	REMARKS
9	G	108.50	107.50	1.0	97.25	10.25	11.25	0.9			Soaked 70 hours.
9	H	106.75	103.0	1.75	91.75	13.25	15.0	1.6			"
9	I	105.75	105.25	0.50	95.0	10.25	10.75	0.5			No. 9, round corners, glazed.
9	J	106.0	105.50	0.50	89.0	16.50	[43.75]	0.5	[18.19]	[18.91]	Broken, small piece, 26.75 oz.
10	G	109.0	108.25	0.75	98.75	9.50	17.0	0.7	11.86	12.61	Soaked 70 hours.
10	H	110.25	109.50	0.75	92.50	17.0	10.25	0.7			"
10	I	103.0	104.0	1.0	87.0	17.0	17.75	0.9			No. 10, round corners. "
11		105.75	105.0	0.75	[50.75]	[54.25]	18.0	0.7	[22.90]	[23.49]	Broken, small piece, 44.5 oz.
11	G	126.0	123.75	2.25	95.25	9.75	[55.0]	1.8	12.48	13.14	Soaked 70 hours. "
11	H	144.0	142.75	1.25	113.50	8.25	10.50	0.9			
11	I	127.75	127.0	0.75	121.75	[78.75]	10.50	0.9			Broken, small piece, 57.75 oz.
11	J	126.50	128.0	0.50	115.50	21.0	[80.0]	0.4	[21.84]	[22.60]	No. 11, round corners, glazed.
12	G	129.50	126.0	1.50	110.0	11.50	22.25	1.2	10.73	11.54	Soaked 70 hours.
12	H	127.50	126.50	1.0	94.25	15.50	12.25	0.8			"
12	I	126.25	121.75	1.50	92.0	29.75	15.50	1.2			No. 12, round corners, grooved transversely.
13	G	133.75	131.0	1.0	98.50	34.50	31.25	2.8	25.18	25.99	Soaked 70 hours.
13	H	131.50	132.75	0.75	114.75	31.25	33.25	2.6			"
13	I	132.25	133.75	0.75	114.0	18.25	35.50	2.6			No. 13, round corners, glazed.
14	G	107.75	132.0	0.75	116.0	18.0	32.75	0.7	14.72	15.24	Soaked 70 hours.
14	H	107.25	107.0	0.25	95.25	17.50	19.0	0.6			"
14	I	105.25	106.50	0.75	95.25	21.50	19.0	0.7			No. 14, round corners.
15	G	103.50	103.0	0.75	96.25	11.75	18.25	0.7	10.44	10.97	Soaked 70 hours.
15	H	76.50	102.75	0.25	92.0	10.75	25.25	0.6			"
15	I	76.25	75.25	1.25	72.25	3.0	12.50	1.6	6.46	[14.91]	Broken in first test, small piece, 21 ounces.
16	I	76.50	[33.50]	[23.50]	70.75	4.75	14.50	1.0		8.06	One specimen only. Round corners, glazed.
16	J	77.75	74.0	2.50	69.50	7.25	[28.0]	3.1	9.71	10.27	
		121.75	121.0	0.75	109.25	11.75	12.50	0.6			

Figures in brackets include weights of pieces which have been saved, as noted in margin.

TABLE No. 6.—COMPARATIVE RANK.

RANK.	ABSORPTION CUBES.	ABSORPTION WHOLE BRICKS.	CRUSHING STRENGTH.	TRANSVERSE STRENGTH.	ABRASION AND IMPACT.	SPECIFIC GRAVITY.
First.......................	13	1 & 15	16	5	1	1
Second....................	1	13	1	7	4	16
Third	15	10	15	6	16	3
Fourth	10	7	14	14	6	7
Fifth.......................	5	6	4	10	14	5-10 & 15
Sixth	7	12	7	15	5	12
Seventh...................	3	3	10	4	7	9
Eighth	9	14	5	3	15	11
Ninth......................	16	16	6	8	13	4 & 6
Tenth......................	11	8	13	13	8	2
Eleventh	12	11	8	9	9	14
Twelfth	8	9	9	12	11	13
Thirteenth	4	5	2	16	10	8
Fourteenth	14	4	3	1	3	....
Fifteenth..................	6	2	11	2	12	....
Sixteenth.................	2	....	2	11	2	....

TABLE No. 7—Loss of Brick in Terms of Granite.

Specimen Number.	First Rattler Test.	Second Rattler Test.	Entire Rattler Test.
1	1.0	1.0	1.0
2	8.1	5.0	5.1
3	3.8	5.1	4.9
4	1.9	2.0	2.0
5	3.0	2.1	2.2
6	2.6	2.1	2.1
7	3.0	2.7	2.7
8	2.6	3.4	3.3
9	2.0	3.8	3.7
10	1.7	4.8	4.5
11	2.1	4.6	4.3
12	2.5	5.3	5.0
13	1.4	3.1	2.9
14	1.4	2.2	2.1
15	20.0	1.4	2.9
16	1.4	2.0	2.0

TABLE No. 8.—CHEMICAL ANALYSES (COMPILED).

NAME OF CLAY.	SILICA.	ALUMINA.	IRON OXIDES.	LIME.	MAGNESIA.	ALKALIES.	WATER.
I Stourbridge Glasshouse Pot Clay	73.82	15.88	2.94	....	....	0.90	6.45
II "Best Clay," Tannworth	71.40	21.17	0.91	0.04	....	0.82	6.06
III Firebrick Clay, Newcastle	55.50	27.75	2.01	0.67	0.73	2.63	10.53
IV Clay from Glasgow	66.16	22.54	5.31	1.42	....	....	3.14
V White Clay, Dorset	65.49	21.28	1.26	4.72	....	7.25	Omitted.
VI Beacon Hill Clay	53.52	33.68	0.52	0.76	0.14	0.04	11.34
VII Hale Paving Brick	66.30	19.53	4.07	1.87	0.42	....	7.24
VIII Mingo Fire Clay	72.24	16.87	0.16	0.50	....	1.09	5.14
IX Burlington, Ia, Brick	77.40	11.74	3.29	1.60	1.91	3.76	Omitted.
X Bloomington, Ia	67.80	11.55	4.31	8.90	5.32	2.42	0.20
XI Slate Brick	56.78	17.38	6.50	4.14	3.15	3.43	7.60
XII Clinton, Iowa	75.50	13.40	4.00	0.50	....	4.50	3.00
XIII Porter Fire Brick	65.60	27.20	5.95	0.48	0.69	0.08	None.
XIV Ground Clay (for same)	60.00	24.84	4.77	0.51	0.73	0.08	8.27

DESCRIPTION OF THE TABLES.

Table No. 1 requires but little description. It shows the essential chemical ingredients of the different varieties of bricks tested as obtained from a single analysis of each kind. Generally the sample was taken from the crushed cubes, but where we failed to crush them pieces were taken from other broken specimens, using the common mortar and pestle for pulverizing. Could a larger quantity of each variety have been ground and used in sampling, or a greater number of analyses have been made, a more accurate average determination would have been obtained, but the time and expense required for such work is so great that it was deemed prudent to limit our investigation to the single analysis. It is evident that an inquiry into the nature of the material of which the bricks are composed is the proper basis for a study of the whole matter, and should next be followed by a scrutiny of the methods of combination. These searches should then be followed by tests of the goods produced. Without proper clay no good result can be expected, and with suitable materials improperly combined failure is sure to follow. From the first cause I would cite numbers 3 and 15, as not meeting the requirements; from the second, numbers 2, 11, and 12 are conspicuous examples.

Could our analyses have been made from the clays used, we should have been working to better advantage; but that was not practicable for us under the circumstances. In order to show that the results obtained are entitled to credence, I have compiled Table No. 8 from the sources named in its description, which shows that the work done has been carefully performed.

Table No. 2, is a detailed statement of the results of the investigations to determine ratios of absorption and specific gravities. The method of performing the work has been described. Numbering the columns from left to right, the first is the specimen number; the second is the weight in grammes of the cubes lettered "A" after being dried; the third is their weight after soaking twelve hours in water; the fourth is the excess of the third over the second, which is obviously the weight of moisture absorbed; the fifth is the quotient (multiplied by one hundred) of the fourth divided by the second, or the percentage of absorption; the sixth is the weight of the cube in water; and the seventh is the specific gravity. This was determined by the formula:

$$\text{Specific gravity} = \frac{W}{W'} \frac{}{W''}$$

In which W equals weight before immerson, W' equals weight after immersion, and W'' equals weight in water. The remainder of the table details similar experiments with whole bricks, using a less accurate means of determining weights, column eight giving the letters designating the marking of specimens of the several numbers (it will be noted that these bricks appear again in The Rattler Test—Table No. 5); column nine giving the weight in ounces before immersion; column ten, the weight after soaking seventy hours in water; column eleven is the excess of ten over nine, being the weight of water absorbed; column twelve is again the percentage, or the quotient multiplied by one hundred of eleven divided by nine; and column thirteen is the mean of the two results from specimens of the same number in column twelve.

The latter test is not one of great accuracy, and the tabulated results can not show all of the influences which should be taken into consideration. For example, numbers three and fifteen were smooth, clean bricks when put in water, but when they were taken out their surfaces were defaced by numerous indentations caused by the dissolution of the contained material. This loss lessens the apparent amount of absorption, and unless provision is made for determining it in each case it can not be measured. It, however, reveals a serious defect, which should not be overlooked. Where nothing of this kind is apparent the test is a very practical one, and can be repeated at pleasure with the ordinary counter-scale and but little trouble. It furnishes a very fair test of the absorbing qualities of the material.

Table No. 3 sets out in detail the results of the work done to ascertain the crushing strength of the materials tested. The method of preparing the specimens having already been described, it is not considered necessary to set forth in further detail the manner of arriving at the results here tabulated, as that will be evident from a mere inspection of the table. A glance at the results obtained shows us that we are not dealing with the substance ordinarily known as brick.

In Table No. 4 is given the work done to ascertain the transverse strength of the material. The Riehle testing machine of Messrs. Otten & Westenhoff was used for this purpose. Three whole bricks or blocks, lettered D, E, and F, of each variety were broken, excepting numbers one and sixteen, of which but one each could conveniently be had. The bricks were supported on knife edges six inches apart, and the load was applied by another knife edge brought to bear midway between, and parallel to, the other two; each point of bearing being protected from direct abrasion by two or three thicknesses of blotting paper.

The modulus of rupture was computed by the ordinary formula:

$$R = \frac{3\,W\,l}{3\,b\,d^4}$$

in which W represents the breaking weight in pounds, b, d,

and l, the breadth, depth, and length, respectively, all in inches, and R the modulus of rupture in pounds. If the span l be measured in feet, while b and d are measured in inches, the formula becomes

$$R = 18 \frac{W\, l}{b\, d^2}$$

Hence, the modulus of rupture is stated by Prof. Rankine to be "Eighteen times the load required to break a bar of one inch square, separated at two points, one foot apart, and loaded in the middle between the points of support." While this is the ordinary formula used in the text-books, and identical with that adopted by Prof. Baker in his recent publication upon "The Durability of Brick Pavements," it should not be confounded with that commonly employed for determining the transverse strength of material, which is

$$R = \frac{l\, W}{4\, a\, d}$$

In which R represents the coefficient of transverse rupture; W the breaking weight; a the sectional area; d the depth; l the length, all in inches. Results obtained by the latter formula will be found to be about ⅛ of those derived from the one first stated.

In the table, the first column on the left gives the specimen numbers; the second, the letters by which they were designated; the third column is the breadth or thickness; the fourth, the vertical depth; the fifth, the length between supports, all of these dimensions being in inches. The sixth column is the product of the third by the fourth, being the area in square inches. The seventh is the weight in pounds at which the specimen was broken; the pressure being exerted by the continuous working of the pump without shock until rupture was produced. The eighth column is the modulus of rupture calculated by the formula first above given. This formula is stated in Professor Baker's pamphlet, page 5, in the following form:

$$W = \frac{2\,b\,d^2}{3\,l} R$$

The letters having the same significance as above given, but a moment's inspection shows this expression to be identical with that used in calculating the table. The ninth column shows the averages of the three results given in column eight for specimens of the same kind or number.

Table No. 5 is a detailed statement of "The Rattler Test," or the effort to measure the effect of abrasion and impact upon the specimens submitted. The method of conducting the test has already been briefly outlined. Tabulating the result has been a tedious process, but it has been done with much care, and is believed to clearly show the results obtained in such a manner that the value of the test can be judged from a correct basis. All the weights were carefully repeated, and, if errors have been made in the calculations, all the work is given in detail, so that corrections can be made by inquiring minds if incorrect results are found in any of the columns.

Counting from the left, the first column gives the specimen number; the second, its letter (those marked G and H having just come from their bath in the absorption test— Table No. 2); the third, its weight in ounces when placed in the rattler; the fourth, its weight after the first thousand revolutions, or at the end of the first test. There was an interval of nearly forty-eight hours between the first and second rattler tests, and the weighing was repeated before placing the specimens the second time in the cleaner, but the

loss in weight by evaporation from the saturated bricks did
not appear to be a measurable quantity by the instrument
used, which was a new Fairbank counter scale, weighing to
quarter ounces. The fifth column is the excess in the
weights given in the third over those in the fourth, or the
loss in ounces of each specimen in the first test. The sixth
column gives the weight of each specimen at the end of the
second or final test. In a few cases, there was uncertainty
about the identity of some specimens, as the marks had been
so defaced, but in all instances the varieties could readily be
distinguished. Where figures are inclosed in parentheses,
they include the weights of the pieces which had been broken
off and were of sufficient size to be saved and weighed.
The seventh column is the excess of the weights given in the
fourth over those in the sixth, being the loss in the second
test. The eighth is the sum of the seventh and fifth, or the
difference of the third and sixth, being the total loss in both
tests. The ninth is the quotient, multiplied by 100, of the
fifth divided by the third, or the percentage, to the nearest
tenth. The tenth column is the percentage of loss in the
second test, and is obtained by dividing one hundred times
the sum of the losses for specimens of the same number,
taken from column seven, by the sum of the weights of
specimens of that number, taken from column four. The
eleventh column is calculated in the same manner, using the
sum of the weights for each variety or number as given in
column eight, and the sum of the weights of the same variety
in column three.

Table No. 6 is compiled from the results set out in the
preceding five tables; the several numbers being rated with
each other in the order in which they have withstood the
various tests to which they have been subjected. It shows
some rather unexpected results and is worthy of very care-

ful study. The rank is by averaging all the tests of each
kind for each number or variety in each test. Thus, in the
absorption tests, those numbers showing a less average percentage of absorption are ranked higher than those showing
a greater percentage. In crushing strength, those showing
a higher power of resistance rank above those showing a
less endurance. In this test, unfortunately, our machine
had not the power to enable us to properly classify the better
grades of material tested, but the rank so given is strictly in
accordance with the results of the work. It is not thought
that No. 16 is absolutely a stronger material than granite,
but one of the granite cubes was slightly spalled at a lower
pressure than was exerted when the No. 16 cube was spalled,
but none of the cubes of either number were crushed, hence
the actual endurance of the material remained undetermined.
In transverse strength, the numbers having a greater average
modulus of rupture are ranked above those having a less,
which correctly classifies the material as to its tenacity under
stress as it was applied in making tests, but furnishes but a
poor criterion by which to judge of the quality of brittleness
under percussion or shocks. Under abrasion and impact,
the numbers are ranked according to the percentages of loss
in the second rattler test; those suffering a less loss being
ranked higher than those suffering a greater one.

Table No. 7 is deduced from the percentages given in
table No. 5, the loss of the granite being taken as one. The
left hand column gives the specimen number; the second
column its ratio of loss in the first test; the third column its
ratio of loss in the second test, and the fourth column its
total ratio of loss in both tests. For purposes of comparison
it is recommended that the figures in the third column be
used. It will be seen that the best bricks under the most

3

favorable conditions suffer about double the loss which the granite does.

Table No. 8. This table has been compiled for the purpose of comparing the work of other investigations with that herein described. The first analysis given is a Stourbridge fire-clay used exclusively for the manufacture of glass house pots and furnace linings. It is a coal measure clay and probably contained traces of lime and magnesia, but no mention of such ingredients is found in Mr. Wills's analysis. The second analysis by the same chemist is of a clay from a like source, but in this case he has determined the percentage of lime contained, which is very small indeed. These clays are mined and used because of their heat resisting qualities, and are principally valuable because at white heat they do not readily vitrify, but retain their form and texture. The third analysis is of a Newcastle fire-clay, by Mr. Taylor, the product being less remarkable for resisting heat than wear. It is to be observed that this clay contains ingredients which at a white heat will unite or vitrify, but would hardly be likely to become fluid or even sufficiently plastic to greatly change in form. It is quite similar to our Nos. 4, 8, and 14, except that it contains much less iron. The fourth analysis, from Percy's Metallurgy, is of the Glasgow fire-clay, adapted to a variety of metallurgical uses, is an average of many determinations in which, unfortunately for our purposes, the percentage of alkalies is not given. With this exception it gives the characteristic ingredients of the coal measure fire-clays. The fifth analysis is of the white clay of the tertiary formation in Dorset, by Professor Way. It is used for the manufacture of fire-brick and could properly be termed a modified fire-clay. The analysis gives the alkalies as "alkalies and alkaline earth," and the lime as a sulphate. This clay contracts very greatly in the process of

drying and burning, to counteract which tendency it is cus-
tomary to incorporate with it fine sand and ground burnt
clay. Number six is the Beacon Hill clay from the Lower
Bagshot Beds, which withstands the high temperatures of
furnace linings without much tendency toward vitrification,
but decrepitates or is worn away by passing currents. The
seventh analysis is taken from a recent publication by C. P.
Chase, on " Brick Pavements," and gives the composition of
the clay used in the manufacture of the Hale paving brick.
Mr. Chase gives the moisture at 212 degrees as 2.08, and
combined water as 5.16, but does not determine the alkalies
present, if any were contained in his sample. If the writer
understands what is meant by the Hale paving brick, this
should correspond with our No. 2, Mr. Chase's analysis having been made from the clay and ours from the brick, but
the resemblance is not very close. The eighth analysis is
copied from the same source as the preceding one. The material in its natural position more nearly resembles a fine-
grained sandstone than a bed of fire-clay, but it pulverizes
readily on exposure to the atmosphere, and its composition
and position have given it its name. From it is manufactured the Hayden block, which is, in reality, a tile used for
paving streets. No sample of this material was furnished
for testing, but it has been used extensively as a paver, and
in some localities is deservedly popular. When properly
burned and annealed it presents a homogeneous, compact
texture, and has great hardness without brittleness. Many
persons would say that the material was " perfectly vitrified,"
but that expression needs specific explanation to be at all in-
telligible. The ninth, tenth, and twelfth analyses are copied
from the same author, who also gives the specific gravity,
crushing strength, and percentage of absorption of the bricks
manufactured from these clays. While he classes them

among " our best paving brick," he gives no statistics showing their enduring qualities in actual use. They would more
nearly coincide with our No. 3 than any other sample tested,
although they might not resemble it in color. The writer
would not regard No. 10 as a very promising composition,
but all of these clays can be melted or vitrified very readily
so as to present a compact texture that will not absorb moisture in any considerable amount. The eleventh analysis is
from the same author of a clay used by the American Brick
and Tile Company, of Phillipsburg, New Jersey. This composition is also said to contain sulphur 0.89, and phosphoric
acid 0.13. No further information is given regarding the
product except that the crushing strength averages from
7,000 to 7,500 pounds per square inch. The thirteenth and
fourteenth analyses were made by Otto Wuth, of Pittsburg,
the first being of the Porter fire-brick and the second of the
ground clay from which such bricks are manufactured.
These compositions, it will be noticed, are quite similar, except that the brick has had the moisture driven off in burning. They closely resemble our No. 14, except that we find
a much higher percentage of the alkalies, and herein lies
the marked distinction between the fire-clay brick, which is
suitable for furnace lining, and the one adapted to use in
street paving.

THE CHEMISTRY OF BRICK MANUFACTURING.

The alkalies of potash in the clays is a residuum of de-
cayed organic matter. It is an active fluxing agent, and in
the process of burning, or so-called vitrification, causes an
amalgamation of the iron and silica components which imparts
a metallic tone or ring to the brick when struck. When
aided by finely pulverized lime or magnesia in the presence
of a large percentage of iron, a pale double silicate of lime

and iron is formed, imparting a buff tint to clays that would otherwise burn red. In the fire-clays less than half of one per cent of potash or alkali produces no noticeable result, and the product has good heat resisting qualities, but when from one to three per cent of this ingredient is found in the clay and it contains from four to eight per cent of iron, which it generally does, with perceptible quantities of lime and magnesia at a high temperature (usually a white heat), these fluxing components form vitreous combinations with the silica, producing a brick quite useless for resisting heat, but when the texture is uniform throughout, and it is allowed to cool gradually, without coming in contact with cold air until below the temperature of boiling water; or, in other words, is properly annealed, you have the so-called vitrified brick, which absorbs about two per cent or less of moisture, and has great strength to resist crushing or abrasion. This product may be used quite fearlessly for street paving.

With the plastic clays or shales the melting or vitrification occurs at a lower temperature, and, owing to the fact that the ingredients are seldom uniformly mingled, there is greater danger of melting the bricks together in the kiln, or of leaving many of them without vitrification. To render them apparently impervious to moisture, many manufacturers have adopted the plan of glazing them with salt, which may be beneficial in some respects, but is objectionable in others. These clays usually contract to a greater extent in the process of drying and burning than the fire-clays do, and hence are more liable to be warped from their proper form, or show injurious fire cracks. But no clay can be made into a good street paving brick, unless the process of firing or burning be continuously progressive and comparatively slow to the maximum temperature, and the cooling down be gradual and continuous. This can not be done in

the ordinary clamp-kiln. A broken brick showing varieties
of texture or color is a certain indication of defective combination or burning, and the fault is fully as liable to be in
the burning as elsewhere. Uniformity in the product of the
kiln is a necessary condition to the successful manufacture
of clay of any kind into proper form to be used for street
paving; and only with such clays, and such appliances as
will enable the manufacturer to attain this result, can he
reasonably hope to achieve success.

DISCUSSION OF THE TESTS.

With the information now before us, what brick shall be
selected? If the tests were of uniform value the numbers
should range in horizontal lines across Table No. 6, the best
material at the top and the poorest at the bottom; but we do
not obtain such results. There are other considerations that
can not appear in the tables. Nearly 50,000 square yards of
surface are to be paved, and the availability must be considered. That which can be promptly furnished in large
quantities should be chosen, even though an article may be
manufactured which is better in some respects, but unattain-
able without great delay. But people whose opinions are
entitled to great respect will honestly differ as to the relative
value of the several tests. For example, Prof. Baker, in his
pamphlet on "Brick Pavements," expresses the opinion decidedly that, "As a test of the quality of brick or stone, the
crushing strength is practically worthless." (Baker on Brick
Pavements, p. 8.) He demonstrates in a concise manner that
the weight on the wheel of a loaded vehicle is not likely to
crush a brick, even though it be soft. Yet the profession generally have considered, and probably will continue to regard it
as essential, that the constructing engineer should be informed
as to the crushing strength of the materials which he uses,

and that, in connection with other information, it is an efficient aid in determining the relative value of different building materials. However, it is only one of the elements to be considered. For example, the crushing strength of cast iron, is about twice that of wrought iron, and of steel more than twice that of cast iron, but this does not make cast iron worth twice, nor steel four times as much as wrought iron for use under compressive stress. In fact, the best practice adopts wrought iron in preference to either of the others for many positions, but a knowledge of the sustaining power of the material is, and must be, essential to the designer. A study of the preceding tables shows that those specimens having a high crushing strength also rank well in the test for abrasions and impact, and it is reasonable to assume that the power to sustain great weight without crushing would be necessary to the durability of a block placed in a roadway, and subjected to the attrition and grinding due to that position. It is true that many experiments or tests are conducted in such a manner, and the results given so incoherently, that they are of little value, but where the work is carefully done, and the record clearly set out, so that knowledge of the comparative strength of different substances can be gained from it, information regarding the crushing strength of any paving material to be used in the form of blocks, will be sought and esteemed as of great merit in determining the value of such material. A recent circular from the State of New York has been placed in the hands of the writer, in which it is stated over a name preceding the title of civil engineer, that the "average resistance to crushing per square inch is 44,000 pounds" for a certain description of brick which had been tested by him. Now, if that civil engineer had informed the public at large by what steps he had arrived at that remarkable conclusion, he would have conferred a benefit upon his fellow

men. The same circular contains further information as follows: " Under an abrasive test equal to a traffic tonnage of
100,000 tons per inch of width, the loss was nine-sixteenths
of an inch, or six and one-fourth per cent of the depth ; thus
under a daily traffic of 100 tons per foot of width of street,
this brick would have a traffic life of twenty-eight years."
That conclusion appears to be quite definite and satisfactory,
but there are ignorant people at large who know neither
just what that abrasive test might be, nor by what process
of reasoning such a result is reached. Some people might
think that a daily traffic of 100 tons per foot of width for a
period of twenty-eight years would not be equal to a traffic
tonnage of 100,000 tons per inch of width, and thus conclude
that, if the first premise is correct, such a pavement would
be good for at least two or three centuries; even upon the
assumption that when it was half worn out the abutting
owners might want it renewed for a change.

The work done at this time for determining crushing
strength is very incomplete, owing to the limited capacity
of the machine, but it is believed to be accurate as far as it
extends, and enables us to properly classify the specimens
not having a resisting capacity exceeding 13,000 pounds per
square inch.

THE ABSORPTION TESTS.

For this class of paving material, a low ratio of absorption is held by many to be a most essential condition, and
therefore that this test is of the greatest importance. In our
work this theory has not been accepted. Of all the specimens tested there is but one (No. 2) which should be rejected
because of its excessive absorption alone, were all other
characteristics satisfactory. Manufacturers have been told
so frequently that a non-absorbent product is a necessary
condition for marketable goods, it is so easy for them to

bring the rate down to two or three per cent, and the test can be so readily made, that but few street paving bricks are in the market which absorb moisture as freely as any of the stone blocks except granite. It is probably unfortunate that no variety of stone, other than Georgia granite, was included in the tests made, but sufficient experiments have been made with the various building stones to show that when the percentage of absorption is three or less, and the material is not laminated, they are neither perceptibly softened, nor made susceptible of destruction by climatic influences. Medina sandstone absorbs from two to four per cent of moisture. Oolitic limestone absorbs from three to five per cent, yet no one asserts that either of these stones is softened or affected detrimentally on this account, and the first is a standard paving stone. Again, of the specimens crushed or subjected to abrasion, there is no indication, unless it be No. 2, in the Rattler test, that any one was weakened by its previous soaking. Therefore, while it is undoubtedly true that a strictly non-absorbent material is the best, yet, among the paving bricks having percentages of absorption lower than three, while the advantage of an exceedingly low rate should not be ignored, other features may be considered. For instance, No. 13 is shown to absorb less moisture than granite, and where it is not to be subjected to an excessive traffic, should on this account be favorably considered, but its endurance under severe tests appears to be exceeded by some of the other varieties.

TRANSVERSE STRENGTH.

The manner in which the transverse strength of the specimens submitted was determined has been described and tabulated, but in doing the work much information was gained that could not be written out. An unexpected de-

gree of strength was exhibited by a majority of the specimens. While this test shows the tenacity of the material
under a stress continually increasing to the point of rupture,
it gives but little information about the ability of the same
substance to withstand the effect of blows or shocks. The
behavior of the bricks at the instant of rupture is instruct-
ive. Some of those which carried the greatest weights were
much shattered. One of the number fives broke into three
triangular pieces of nearly equal size. Nearly every specimen which exhibited the characteristic vitrified appearance,
threw off flint-like spalls, and presented an irregular fracture. Those specimens which in other tests manifested the
greatest endurance were usually parted by a clean fracture
almost at right angles with the brick, directly beneath the
central bearing, like the granite; while those having interior
defects of any kind would separate at any point between the
outer bearings. This test, therefore, is of much value to the
experimenter, but the tabulated result is not a sure indication of the value of the material for street paving purposes.

ABRASION AND IMPACT.

The manner of conducting this work has been so fully
described, and the results set out in such detail in Tables 5
and 7, that further comment is hardly necessary, yet it is
plain that it presents no condition at all similar to that
which obtains in actual service. The bricks are loose and
battered upon all surfaces, whereas in the pavement they
are held firmly in place and subjected to abrasion upon one
side only. But in this case they were all subjected to the
same treatment, and their losses should give a fair measure
of their relative powers of resistance. "The Rattler Test"
has been frequently repeated by various parties, and a practice is coming in vogue of assuming that a half hour or an

hour in the rattler is equivalent to a year's wear in the pavement under a given amount of traffic, and from this assumption the probable life of the brick in actual use in the street is calculated. By a somewhat similar course of reasoning, although the premises are more fully and fairly detailed, Professor Baker has calculated Table No. 7, given on pages 32 and 33 of his pamphlet on Brick Pavements, in which the life of a pavement made of each of the varieties of brick which he tested, is given in certain streets of ten of the principal cities of this country. The daily traffic tonnage is taken from Captain Greene's statistics, and the results as tabulated are remarkable. His poorest brick would last four years on Broadway, New York, and one hundred and sixty-five years on Olive street, St. Louis; while his best brick would last thirty-eight years on Broadway, and fifteen hundred and twenty years on Olive street. The writer does not dispute such conclusions, but has no facts from which similar inferences can be drawn.

Statistics of Traffic and Durabililty of Pavements.

Data regarding the traffic tonnage, and the effect of such wear on street pavements and highways, has not been collected and preserved in this country in such form as to be available for ready reference. A few years since, Captain F. V. Greene prepared a paper, "An Account of Some Observations of Street Traffic," which was published in Volume 15 of the transactions of The American Society of Civil Engineers. The observations were made by employés of The Barber Asphalt Paving Company, under Captain Greene's directions, during the months of October and November, 1885, in the ten large cities in which that company had offices and works. "The agent in each city was instructed to select the three streets in that city paved with stone,

asphalt, and wood (if any existed), which, by common report, had the heaviest traffic in the class of pavement used
on that street. The record was in every case made on six
consecutive days (Sundays omitted), at the same place, and
it was continuous from 7 A. M. to 7 P. M., except when dark-.
ness prevented. No addition was made for this omission ;
no record was kept during the night, and no addition was
made as an estimate of night traffic." "The traffic is divided into three classes, light weight (less than one ton),
medium weight (between one and three tons), and heavy
weight (more than three tons).

The Captain says : " I have discarded the weight of the
horses altogether, not because they do not constitute a factor
in the wear of the pavement, but because they were discarded in the English reports, and I desired, as far as possi-
ble, to make comparisons with them." . . . "To obtain
the tonnage, I estimated the light weight vehicles to average
one-half ton each (including their loads), the medium weight
two tons, and the heavy weight four tons."

LIGHT WEIGHT INCLUDED..... { 1-horse carriages, empty or loaded.
1-horse wagons, empty or light-loaded.
1-horse carts, empty.

MEDIUM WEIGHT INCLUDED.. { 1-horse wagons, heavy-loaded.
1-horse carts, loaded.
2-horse wagons, empty or light-loaded.

HEAVY WEIGHT INCLUDED... { Wagons or trucks drawn by two or more
horses, and heavy loaded.

" The average tonnage per vehicle ranges from 0.68 on
Fifth avenue (New York) to 2.08 on a portion of Wabash
avenue (Chicago). On Fifth avenue, 91 per cent of all the
vehicles weigh less than one ton, while on Wabash avenue,
only 25 per cent of them have so little weight. The general
average for all the cities is as follows : Less than one ton,
67 per cent ; between one and three tons, 26 per cent ; more-

than three tons, 7 per cent. The average tonnage per foot in each city, so far as here observed, varies from 151 in New York to 30 in Buffalo, and the general average is 77. For all the cities in the table, the average daily tonnage per foot of width is 77, and varies from 273 tons on Broadway to 7 tons on a granite street in St. Louis. The average weight per vehicle is, for all the cities, 1.15 tons. The average width of street between curbs is 44 feet."

This is believed to be the first carefully prepared census of travel made public in this country, and it was published by an officer of an asphalt paving company. It is fair to presume that one object in view was to show the durability of that kind of pavement under heavy traffic. Since its publication, a few annual reports have contained statistics upon the subject, and the investigation has been greatly extended by the different asphalt paving companies. Obviously, information of this kind should be officially compiled by municipal officers upon a uniform system throughout the country and its scope materially extended. The effect or wear upon the roadway of an observed traffic tonnage should be given, which has not been done except in a few of the English reports, and there mainly in cost of maintenance or repairs. Reports from Washington have given some data as to the cost of maintenance of certain pavements, and the English reports are usually quite explicit upon this point; but it would greatly benefit all municipal corporations in this country, were each to keep a record of the kind of street improvements made, their manner and cost of construction; their durability and expense of maintenance, under a traffic, the volume of which could be noted with reasonable accuracy, at but trifling expense. The omission of the horses from the traffic census is clearly a fault, as we know that they assist largely in wearing the roadway. For example,

between the rails of street car tracks upon lines operated by
horses or mules, the wear of the pavement is due almost exclusively to this cause, and it is known to be very great.
The tonnage of vehicles, as estimated by Captain Greene, is
heavier than many observers would assume it to be, and the
percentage to be added for the weight of animals will vary
with the nature of the traffic, being greater with the light
and less with the heavy traffic. His estimate being, that on
Fifth avenue, which carries 91 per cent light traffic, the addition should be about 85 per cent for the horses, while on
Wabash avenue, where but 25 per cent is light traffic, the
addition should be only about 40 per cent. The effect of the
horse's shoe upon the street surface is modified by the nature
of the pavement. Probably sheet asphalt suffers as little
from it as any known form of wearing surface, unless the
blows fall successively upon the same place and thus effect a
displacement of the material. The bowlder is seldom scarred
by it, hence the material of the cobble stone pavement is
practically indestructible from this cause. Granite blocks
are spalled and rounded until they assume the form of bowl-
ders, and, if very hard, become exceedingly slippery and afford insecure footing. Brick pavements would be rapidly
destroyed were the bricks as widely separated as granite
blocks usually are, but being placed in close contact, there is
little room for the rounding away of corners. The brick
surface is, however, affected as it would be by receiving a
like blow from a cutting tool or chisel of similar form.
From this cause, will result by far the greater portion of the
wear, since the pavement, when unbroken, will be sufficiently
smooth to present but slight obstacles to the rolling upon it
of the wheels of vehicles, and it will suffer comparatively
little from that cause. The blow delivered by the animal's
shoe will be greatly increased at high speeds. It would,

therefore, appear to be proper, that upon avenues carrying
suburban travel, a census of traffic should take cognizance of
the element of speed.

THE PROBABLE DURABILITY OF A BRICK PAVEMENT.

This chipping or abrasion of the surface by the shoes of
animals traveling upon it will be its severest trial, and since
no definite statistics are available by which to compute the
traffic tonnage to which it will be subjected, and no test has
been made which serves as an actual measure of the wear
of a pavement under a given tonnage, the probable durability of this street can not be stated, but can only be predi-
cated upon the endurance of the brick as compared with the
granite. Judging this street by others upon which the census of travel has been taken, it seems fair to assume that the
traffic will not greatly exceed 60 tons per foot of width per
day, including the weight of horses, which will probably
embrace one-half of it. The surface of a granite block
pavement, as ordinarily constructed, is about 75 per cent
granite, while a brick pavement is about 90 per cent brick.
There is, therefore, about 20 per cent more brick than granite to resist wear. The brick surface is comparatively
smooth, while the granite is uneven. Wheels will roll
smoothly over the brick, while they will jolt over the granite
with a continual succession of blows. Let it be assumed
that the wear due to horses on the brick will be 120 per cent
of that due to the same cause on the granite, and the wear
due to vehicles on the granite is 200 per cent of that due to
the same cause on the brick: it follows that the total effect on the brick is but 80 per cent of that on the granite.
Now we find in table No. 7 the loss of the brick in our
abrasion test to be 2.2 times that of the granite; a traffic, therefore, which wears off one inch from the granite pavement

will wear one and two-thirds inches from the brick; or, the time required to wear an inch from the brick will be about 60 per cent of that required to wear an inch from the granite. No record is known to exist showing that amount of wear from a granite block pavement under a similar traffic, but about five times the tonnage has worn some portions of our city pavements to about that depth in four years. The estimated traffic is about 60 per cent of that on Fourth street between Walnut and Race streets, excluding street cars, and quite similar in character, taking the entire width of pavement (omitting car tracks), and five years' wear has been estimated to have reduced the blocks one-fourth of an inch. This would seem to justify the belief that this pavement should be in fair condition after ten years' traffic shall have passed over it.

MUNICIPAL METHODS.

A cause for the lack of definite statistics upon these matters is apparent when municipal methods are considered. American civil engineers have achieved a world-wide reputation for the boldness and originality of their designs, the skill exhibited in their execution, and the economy shown in attaining results. Great industrial establishments have been built, lines of transportation, with all the works appertaining thereto, have been by them located and constructed, and they are accredited with being well toward the van, and of contributing their full share toward the progress and development of the country. In all such works facts have been collected and compiled, so that reliable data is available. Manufacturers are willing to guarantee a given mileage for their steel rails or car wheels, or a given strength for their iron and steel, from data made available by engineers, but in municipal matters the conditions or the results are in no way

similar. The total amount of money annually expended by
the municipalities of the country in opening, improving,
cleaning, and repairing streets and highways, is an enormous
sum, exceeding that applied upon all other public works in
an equal length of time. The greater part of this fund is
nominally disbursed under the supervision of engineers, but
the results are not such as to add materially to the renown
of the profession, or to supply exact data for their guidance
in present or future works of this character. One reason for
this appears to be found in the fact that these funds furnish
the greatest of the existing causes of activity in local pol-
itics. Municipal statesmanship is developed in levying,
watching, and disbursing this money. Laws providing for
its collection, control, and disbursement have been enacted,
termed the municipal code, which is more complex and pe-
culiar than any other system known to man. Boards and Bureaus, Councils and Commissioners, Supervisors and Direct-
ors, Counselors and Barristers, have been created or called to
govern the work, guard the public interest, acquire fame, and
enjoy the advantages accruing to exalted official position. The
pervading spirit of freedom abroad in the land being averse to
the creation of such a class as controls similar matters on
the continent of Europe, the rights of the people are sought
to be preserved by the checking and balancing of sovereign
and independent departments. When new things are to be
done, additional statutes are enacted and more boards provided. One authority will make an improvement and another will dig it up, while no one will repair it because the
courts have not decided the question as to which fund shall
be drawn upon for meeting the expense in cases of that nature. Volumes of annual reports from the heads of the sovereign departments and chiefs of the multitudinous bureaus
into which they have been sub-divided, assure a confiding

4

public, that, since the advent to power of the present incum-
bent, the affairs under his control have been conducted upon
strictly business principles, thus enabling him to grant more
permits and file a larger number of papers than had ever before been handled by similar officers in a corresponding
length of time. When the balance of power between political parties is indefinite, and changes in official stations
become so frequent as to make employment uncertain, it
is sometimes found expedient to further revise the statutes and make non-partisan boards, who then carefully di-
vide the appointments and perquisites between contending
parties, accurate data for the making of such partitions being always available. The smaller municipal organizations
copy the " systems" of the larger ones.

Under such regulations no very considerable amount of
" engineering " is required. A " chief engineer " of suitable
political complexion is chosen to sign the necessary papers,
to whom matters not well understood by other parties can be
referred and reported upon, and who can be blamed when it
becomes absolutely necessary to locate responsibility some
place, and who is willing to allow officials and other influ-
ential parties to appoint his assistants, clerks, rodmen, and
superintendents. Men who have acquired skill and experience in the construction of works under different regulations
seldom take kindly to this order of things, and the field is left
free to such as enjoy the surroundings. Many careful and
painstaking men are engaged in city work who would make
excellent records were they not handicapped by the regulations governing them, and almost the entire number are
like the parents of heroes, " poor but respectable." Having
little at stake except their integrity, that is manfully cher-
ished. Occasionally an erring brother may fall, but he
merely drops from the ranks which close in his place. The

ammunition of the enemy, which is most dangerous, especially to those of limited experience, is flattery. Not one
person in ten thousand of those having experience upon
public works would ever approach an engineer with money
or a valuable consideration for corrupt purposes, but if the
insidious agent can induce him to believe that his genius is
apparent to all, and that the world, especially the official
part of it, will soon be shouting his praises, such influences
may cause the young man to make himself ridiculous. But
there is a great following who have a sufficient knowledge
of surveying to enable them to handle field instruments, set
out work, and compute quantities, who have but little taste
for such study or investigation as is necessary to acquaint
them with materials, or render them skillful in designing or
constructing engineering works; and their appears to be a
greater demand for these persons on municipal, than upon any
other class of public works. This is probably because they
have more leisure for compiling political statistics than
others, are less liable to have bothersome opinions about
how things should be done, and can more readily discern
the grade and character of improvements desired by those *in*
power or opposed by those *not* in power, which is usually the
same thing; it being always understood among municipal
statesmen that an election or appointment to office confers
upon the recipient of such honor all the necessary knowledge and experience required, not only to choose an engineer,
but to tell him what to do, and just how to do it. Many re-
cruits are obtained from those estimable young men annually graduated from our technical schools and colleges. The
learned professors solemnly announce to such of their stu-
dents as have pursued certain lines of study, that they are
now civil engineers. The young men very properly have
great confidence in their teachers, and believe what is told

them to be literally true, but when they go abroad in the
world and learn that what the professors really meant was,
that they were qualified to obtain employment upon public
works, where, by continual study and actual practice, they
could become engineers; the shock is very great, many never
recover, and some are engaged by municipal corporations.
Such as do recover are achieving great success in professional
work.

The number and sovereignty of the departments, the
uncertainty of the laws (for no one dare hazard any thing
more than an *opinion* regarding the rule of action prescribed
by a statute until the court of last resort has guessed at its
meaning), and the strifes of contending factions have produced conditions so different from those which would appear to be proper that heroic measures may be required to
effect desirable changes. Unless the people at large can be
induced to look upon the matter of municipal government as
a grotesque absurdity which is really being enacted at their
expense, as it is, the code will continue to be enlarged and the
Boards multiplied. When they induce the law-makers to re-
peal the thousand and one statutes which now exist, and
enact a plain, concise code of rules, *and not amend it*, which
will place the direction of public works under a single department, with uniform regulations in like municipalities
throughout the state; placing the designing and manage-
ment in the hands of a corps of engineers who should acquire position and promotion by the record of their achieve-
ments, and not by reason of race, creed, or previous condition of partisan servitude or influence, and who, being un-
trammeled as in the world at large, would succeed or fail by
merit alone, the principal of natural selection, or the survival
of the fittest, would soon place the direction of such works
in systematic order under competent control. Then would

streets be built to remain undisturbed, as the bottom layers or drains and pipes would first be put down and carried to property lines, then would the character of the pavement be adapted to the uses to which it would be subjected. Paving companies would construct streets and guarantee them to remain in proper form and repair until a specified traffic tonnage should have passed over them. Manufacturers would furnish materials under like conditions. Order and uniform system would exist where chaos now reigns, and legislative interference would cease to trouble executive business.

Probably the view is Utopian, and will never be realized until we pace those golden streets, but the patching remedies and special laws continually being enacted for the betterment of evils known to exist are only adding complications to complex affairs, and if thinking people are induced to direct their attention to a subject of such universal and vital interest, and make an effort in unison to better municipal government as applied to public works it will certainly result in some good. Politicians and bosses will undoubtedly offer great obstacles, but the mere absurdity of present methods will insure a change, and if engineers were accorded similar freedom and control, with such responsibilities and opportunities as are given them upon other works, they would not ignore so inviting a field as that presented by the needed improvements in these matters.

GENERAL DISCUSSION OF PAVEMENTS.

The office of a street pavement is to provide a wearing surface which shall fulfill the following conditions :

First. It must present a secure and pleasant footing for animals.

Second. It must have sufficient smoothness to render traveling in carriages agreeable, and traction easy and as nearly

noiseless as is practicable, for all descriptions of wheeled vehicles (excepting those provided with *flanged* wheels).

Third. It must be of such form and material that liquids
falling upon it will quickly flow from it into proper conduits,
and must furnish no permanent lodgment for street filth of
any kind.

Fourth. It must be capable of sustaining without change
of form, any and all loads usually transported on public
highways.

Fifth. It must be reasonably durable, both as against
the attrition of street traffic, and the destroying or dissolving
action of the elements.

Sixth. It must be economical. That is to say, sufficient
comfortable use must be obtained from it to make it worth
both the cost of construction and maintenance.

Seventh. It must be capable of removal and replace-
ment, or repair from failure at reasonable cost, and with
materials and appliances within the control of the street repairing department.

A study of these conditions at once reveals the reason
why the " paving problem " is of such an intricate nature
that it has so long remained unsolved, as well as a cause for
so many unhappy failures in its attempted solution.

For the first and second conditions, the dirt road in good
repair stands without a rival, but it meets no other requirement, hence its use is restricted to race tracks and country
roads, which like canals are only navigable when the weather
conditions are favorable.

For the second, third, and fourth conditions, the asphalt
pavement on proper foundation appears to be better fitted
than any other that has come into such general use; but
many persons say that it does not properly meet the first re-

quirement, criticise it severely as to the fifth and sixth, and affirm that it utterly fails to meet the seventh.

Stone block pavements meet the first requirement, but indifferently; utterly fail in the second and third, when properly constructed; are better adapted to comply with the conditions of the fourth, fifth, and seventh, than almost any other description of city street, but when a high charge for transportation is to be added to the cost of preparing the material, they fail to meet the sixth condition.

Wooden block pavements meet the first, second, fourth, and seventh conditions fairly well, when made of suitable materials well combined; but, as they have been built in this country, have signally failed to meet the third condition, and have fulfilled the fifth and sixth but very indifferently.

The bowlder or cobble-stone pavement has been with us so long, and has been treated so badly, that familiarity with it has bred a species of contempt that is hard to overcome. It has become popular to consider it an all around failure, yet it meets the first and seventh conditions fairly well, and so far as the material is concerned, it stands unrivaled in the fifth. In many of our cities where horse cars have been operated for the past twenty or thirty years, and the street railway companies are required to maintain the pavements within their tracks, the bowlder pavements are still retained between the rails, while the residue of the streets have been paved with other materials, because in that position they are *said* to meet all of the conditions named, excepting possibly the second and third, better than any other substance yet offered for the wearing surface of roadways. This *saying*, however, does not appear to be any thing more than an expression of opinion, which can not be sustained by any process of reasoning. The cobble-stone can be given no definite bearing on any foundation; it can not be held in position by any bond

that can be given it in construction. It does not present a
suitable surface for vehicular travel, or that can by any process be kept free from filth; yet it does not wear out, is easily
restored if loosened from its place, and it does answer very
well for street car horses to travel upon.

Broken stone or macadam as commonly used, of mingled
limestone and shale, meets none of the requirements. If,
however, it is clean refractory material, properly prepared
and combined by rolling, it fulfills all the conditions except
the third (and even that reasonably well), providing the traffic
is moderate, and the repairing is promptly and efficiently
done. It may be set down as an established fact, however,
that when a macadamized street is dug into for any purpose
that it is never properly replaced.

No one of these conditions can be entirely ignored, yet
it is obvious that no pavement yet devised, fully meets all of
them. Could the first be ignored, it would be an easy matter to cover street surfaces with iron or steel plates that
would fully meet all the others, but plainly this can not be
done. The surroundings of the pavement and the extent and
nature of the traffic to which it is to be subjected, must be
considered in order to decide which of the conditions
shall be allowed to determine its character. The first, that
of furnishing a secure and reasonably comfortable footing
for animals, can in no case be ignored, and in many instances
must control all other considerations. Wherever the pavement is to be used as a thoroughfare for vehicular traffic at
fair rates of speed, or when time, pleasure driving, or quiet-
ness become elements of importance, then the first and second conditions must be met, and other features may or may
not be caused to yield to their requirements. But the preservation of life and health is the essential cause of business

activity, hence the third condition, that of maintaining correct sanitary conditions, should never be neglected.

To those at all familiar with street construction, it is obvious that the wearing surface, or pavement proper, can not, and does not in itself, support the loads brought upon it, but that it more or less successfully resists the impact and abrasion incident to the traffic, and transmits the weight directly to the bed or foundation upon which this surface material has been placed. It follows, therefore, that the fourth condition can be met by any description of paving material which has sufficient hardness to retain its form under the pressure of street traffic, by merely placing it on a properly prepared foundation; and further, that unless the pavement shall be placed upon a bed capable of sustaining under all conditions the loads brought upon it, the surface will yield regardless of the material of which it is composed, and that this condition not being complied with, no essential feature of a good street surface will remain. Failure to meet this condition is the error most commonly committed in the building of pavements. In this latitude the winter frosts penetrate to a depth of from one to three feet, or, when not acted upon by frost, the subsoil drainage is seldom so thoroughly efficient as to prevent the changing of the ground from a firm unyielding soil to one of almost complete saturation, thus materially affecting its sustaining power. It therefore follows that no pavement which is to be subjected to a heavy traffic at all seasons of the year can be relied upon to retain the form originally given it, unless the foundation or bed upon which it is placed shall either be carried below the action of the frost, say three feet or more, or be so constructed as to distribute the weights of passing loads over sufficient areas to enable a comparatively weak subsoil to sustain · them. The deep foundation is the ancient, and undoubtedly

the most durable method, having apparently been the ordinary practice with the Romans, but the distributing coating
is far more economical, and hence has become the established
modern practice.

Two methods are in vogue. First, to drain the sub-
roadway as efficiently as is practicable, grade it to the proper
form, compact its surface by rolling, and cover it with a
layer of mingled broken stone and gravel, which is made
smooth and firm by flooding and rolling with a steam roller;
the layer of metal being from six to twelve inches in thickness, according to the requirements of the locality or the
specifications. On this layer or "foundation" is spread the
bed of sand, in or upon which the pavement'is set. Sometimes broken stone alone, and again gravel only, is used for
the bottom course. This style of "foundation" is used very
extensively for all descriptions of pavements excepting asphalt. With brick pavements the practice of placing a layer
of bricks flatwise on the bed of sand, covering them with a
thin coating of sand, and paving on it the wearing surface
on edge, is quite common, and produces what is called the
"two course" pavement. Still another method consists in
covering the layer of sand with tarred boards, upon which
the sand cushion and brick on edge are paved herring-bone
style, producing the "Hale Pavement." In this, however,
the broken stone is generally omitted, the boards being separated from the subsoil by from four to six inches of sand
only. These expedients tend to better the distribution
of the weights brought upon the pavement, and have the
merit of economy in first cost, but they are obviously inade-
quate except where the subsoil is exceptionally good and the
traffic very moderate. The method of combination is quite
defective. When gravel is used that is free from loam, it
will not compact under the roller, and if it does contain

loam the water which comes from the subsoil, and percolates through it, is liable to carry the soluble substances with it down the gradients, and leave the pavement unevenly supported. When broken stone and gravel, or broken stone alone, forms the foundation course, it is expected to be porous, and act, to some extent, as a subsoil drain. The voids, however, are liable to become the receptacles of the clay from beneath, which is brought up, or rather the stones brought down, by the pressure upon the pavement, or they will be filled by the sifting down of the bedding course of sand, caused by the jar of the travel, and this escape of the sand will leave the blocks unevenly supported. All of the varieties described in this first method are extensively used, and are made more or less expensive and durable, or cheap and temporary, as they are carried to greater or less depths, and as the work is thoroughly or carelessly done. But they are so constructed that natural causes would alone destroy them in a comparatively brief space of time, and when the forces of nature are aided by the disturbances to which the sub-grade of the street is ordinarily subjected, and the traffic upon the pavement, it follows that the life of such a foundation seldom exceeds the duration of the wearing surface, and the failure of the former very frequently accelerates the destruction of the latter.

The second method consists in preparing the subsoil by grading and rolling as before described, and placing upon it a layer of hydraulic cement concrete to serve as a foundation for the pavement. For equal volumes, the cost of the concrete is about three times that of the broken stone or gravel; but from one-half to two-thirds of the amount is required, hence the expense of the concrete foundation is one and a half to twice that of the broken stone or gravel. When properly made and undisturbed, it will not yield to

the action of the weather, and the renewal of the pavement
need extend to the wearing surface only. The expense of
cutting through and replacing the concrete when the street
must be opened for any purpose, is much greater (perhaps
two to three times) than in the other forms of foundation,
but such work can be done without serious injury to the remainder of the street; and when repairs are properly made
the opening of the pavement and its foundation is less injurious to the street having a concrete foundation than the
one that has it not, because the concrete base will support
the pavement over small cavities, while the broken stone or
gravel will sink into them.

The thickness of the concrete varies with the requirements of the traffic, and other conditions, from four to
eight or more inches. The ordinary practice is to use natural cement in its composition, and make the coating six
inches in thickness for roadways of medium traffic without
car tracks. The condition of the subsoil should, however,
be considered in determining the depth of concrete, for
where it is soft or spongy, or trenches are to be spanned, a
greater amount will be required. A concrete foundation is
an absolutely necessary beginning for any really good and
durable street pavement, and even for work of medium
character and price it is economical. Pavements of sheet
asphalt are always placed on concrete foundations, the wearing surface being separated from the cement by a cushion
coat, ordinarily about half an inch in thickness.

Stone, asphalt, wooden block, or brick pavements, are
usually placed on a layer of sand from one to two inches in
thickness over the concrete, but the practice regarding the
cushion coat is by no means uniform, varying from an actual
bedding of the blocks in the cement mortar to two or even
three inches of sand, but the *general* custom in this country

appears to be in favor of the sand cushion. Convenience in construction and repairs and the theoretical elasticity of surface being in favor of that combination. The choice of the sand to be used is of more than ordinary importance, since if it contains a considerable percentage of soluble substances, or is alternately coarse and fine in different places, displacement is likely to occur, and the pavement will become uneven.

All block pavements, whether of stone, wood, asphalt, or brick, should be as closely placed as is practicable, and the interstices filled with a non-absorbent material. Possibly a narrow spacing may be required on gradients paved with sawed wooden blocks to furnish footing for animals, but the wide spacing ostensibly for this purpose so frequently seen in all these varieties of pavements is undoubtedly bad practice, since it so materially reduces the resistance of the material composing the wearing surface, facilitates the chipping from the angles, and produces an uneven, noisy street, with cavities to receive and retain filth.

What Shall be Specified.

The writer has seen no specifications for brick pavements which either clearly or fairly describe the material to be used, so that manufacturers or bidders can know just what will be required upon a given work. The majority of the specifications recite that "the brick to be used must be hard, free from defects of any kind, manufactured and burned especially for street paving purposes, be equal in all respects to the sample filed with the proposal, and subject to inspection and acceptance or rejection by the engineer or inspector." With our present knowledge of this material, this phraseology may be accepted, but in reality it specifies very little. Its acceptance or rejection by an engineer is ordinarily regarded

by the corporation as a safeguard, but no parties entering
into a contract can place a power that is wholly arbitrary
and undefined in the hands of an engineer, who is in reality
the executive agent of one of the parties. Because bricks
have been manufactured and burned especially for street
paving purposes, does not necessarily fit them for such use;
the term *hard* is an indefinite one, and without stating what
constitutes defects, there may be differences of opinion as to
whether or not they exist in a given article as well as to the
equality of goods furnished with sample exhibited. Even
sample paving bricks have been known to be worthless for
street paving. The characteristic qualities and strength of
the material are not clearly set forth as they should be. The
power to accept or reject is left undefined, which should
never be the case, neither manufacturer, bidder, nor tax-payer
should be bound by the action of an engineer, unless that action shall be in accordance with known provisions and fixed
rules. With the hope, therefore, of adding something to the
information needed for bettering specifications in this respect, let us examine our work as tabulated.

From Tables Nos. 1 and 8, we might select a chemical
composition quite suitable for the purpose, but there are such
wide variations in clays that it will probably be advisable not
to be too definite in specifying ingredients. Silica may constitute from sixty to seventy-five per cent; alumina, from fifteen to twenty-five; iron, from five to ten; while lime and
magnesia should neither exceed two per cent; and the alkalies should be from one to two per cent. The form and
proportions in which these ingredients are present, and the
difference in results obtainable by variations in methods of
manufacture are so great, that it will evidently be better to
define the qualities which are required, and leave the manufacturer free to produce them in his own way. Lime can

be readily detected, and an excess of that ingredient might properly be prohibited by the specification.

With suitable ingredients, properly combined and burned, the percentage of absorption will be small. An inspection of Table No. 2 shows that it varies greatly, indicating a lack of uniformity in results, which is to be avoided, or failure will result. The absorption should, therefore, be placed at the lowest limit which can be designated without excluding such bricks as are known to have withstood other tests creditably. I would recommend naming two per cent as a maximum.

Specific gravity depends to some extent on the density of the material, and should, therefore, be as high as is obtainable with this material. Two and one-tenth might be named as a minimum, with credit for excess, as about two and three-tenths is attainable.

The crushing strength determined from two inch cubes, prepared and treated as set out in describing the tests shown in Table No. 3, should not be less than 12,000 pounds per square inch. This might be named as the average crushing strength, the limit of variation of any sample not to exceed twenty-five per cent below.

The modulus of rupture, as determined by the formula used in calculating Table No. 4, from tests of transverse strength, could properly be fixed at 1600.

Tables No. 5 and 7 would indicate that unless a rather liberal allowance be made in the matter of abrasion as compared with granite, manufacturers may complain that many are called but few are chosen. One-third of the competitors in this test would be among the saved by placing the limit at two and two-tenths times. It will be safe, therefore, to say that in any tests for determining the comparative loss of the brick under abrasion and impact, as compared with Lithonia granite, the loss in weight of the brick shall not be more than two and one-fourth times that of the granite.

Uniformity in size and texture appears to be attainable, and lack of it produces many undesirable difficulties. Wide variations in size result from faulty drying and burning. We must either specify how the drying of the brick and the firing of the kiln shall be done, or describe what we want as the product of the kiln, and insist upon getting it. In the present state of our knowledge the safer plan will be to name the requirements and leave the manufacturer to wrestle with details.

The amount or percentage of contraction in drying and burning varies with the different clays from one to twelve per cent. With the coal measure fire-clays it is usually about two or three per cent, while with the plastic clays and river silt it is very much greater. Excessive contraction is obviated to some extent by adding to the mixture ground burnt clay and a fusible sand. Unless other ingredients are present to obviate such a result, these components are liable to produce brittleness. In modern down-draft kilns with clays which do, and should contract as much as three per cent, rapid firing sometimes produces bricks from the upper courses, on which the flame acts directly, that are actually larger than the molds in which they have been formed. It follows that such bricks are much checked and cracked, rendering them unfit for use, but the cause is apparent. The contained moisture in the partially dried brick is expanded by being first turned into steam, and, while in this condition, the outside of the brick is fused, thus permanently fixing its exterior form and dimensions, except as modified by checks and fire cracks. Other bricks in the kiln, made in the same molds, will be shrunken to the full extent of their contractility, thus producing not only a troublesome variation in size, but likewise wide differences in quality. These defects can be remedied by a gradual and continuous firing that will produce and permit the natural amount of contraction throughout the kiln. Absolute uniformity in size is not practically attainable, but a much nearer approach to it than is now common can be reached by specifying a minimum allowable deviation and rejecting material that does not comply with the requirement. I would suggest two per cent as the maximum allowable variation from the standard dimensions adopted.

Assuming that the clay has been properly ground and mixed, uniformity in texture is obtainable only by conducting the burning in a suitable manner, continuing it to a sufficient degree, and no longer, and then allowing the kiln to cool down without permitting drafts of cold air to come in contact with the bricks until their temperature has fallen below the boiling point of water. The degree to which the firing should be extended is ordinarily termed "complete vitrification," but would appear to the writer to have been very unhappily chosen. To " vitrify " is ordinarily held to signify " turning into glass by the action of heat," which is not the desired result in this case. The term having been adopted by common consent, the meaning which defines the process must be given it, and would appear to be, that the alkalies and alkaline earths shall form vitreous compounds with the iron and more readily fusible silica, so as to give the brick a dense, uniform texture, obliterating the granular appearance completely, but not fusing or melting the brick so as to injure its form or exterior. This result is obtainable only by progressive firing to the right extent, but with the manufacturer it is like Major Jones's exercises, the fine point consists in knowing just where to stop. In this particular feature the fire-clays are claimed to be superior to the shales and

5

silts. The latter are more readily fused than the former, and
when the melting point is reached the bricks sometimes lose
their form and melt together. To avoid danger from this
source sudden cooling is resorted to and the product is rendered practically worthless by the brittleness which results
from such a course. It is claimed that the fire-clays can be
held at a sufficiently high temperature to produce the required " vitrification " without danger of melting together,
and hence furnish the most reliable product for street paving
purposes. Our investigations appear to point to the conclusion, that from neither clay has that uniformity of product
been attained which is desirable, and that can and must be
made before reasonable certainty in strength and durability
can be assured; for among the fire-clay bricks many are
found that show scarcely any indications of fusion or " vitrification " at all. Many more that are but partially fused or
" vitrified," the exterior portion being dense and non-absorbent, while the interior, marked by concentric colored
rings, surrounds a central portion of open granular texture,
and still others present a closed metallic or granitic texture
throughout. By firing and annealing in a proper manner,
uniformity in texture without brittleness can be produced
from any clay that is suitable for the manufacture of bricks
for street paving. These qualities should be specified, and
such manufacturers as can not and do not produce them,
should not find sale for their goods, because the hopeful
young industry will soon perish or be relegated to the smaller
interior cities and towns, unless these essential qualities are
produced. It may be asserted that such requirements will
increase the cost of production. If need be let it be so, but
such as furnish the required product will find a continued
and increasing demand for their goods, giving a permanent
value to the plant, and the only additional expense necessa-

rily involved would appear to be more time and care devoted to the burning and cooling.

The following is suggested as an addition to the specifications, and as more knowledge is acquired, further revision may be necessary.

"The bricks or blocks to be used for paving shall be straight, smooth, and free from checks or fire cracks. The corners shall be rounded to a radius of one-fourth of an inch. In size they shall not vary more than two per cent in any dimension from the standard adopted for the kind of bricks or blocks to be used. When broken, the fracture shall be smooth and straight, not conchoidal; and the texture of the block shall be uniform throughout, and not granular. The amount of moisture absorbed when tests are made either with the whole block or pieces, shall not exceed two per cent of the weight of any sample when continuously immersed for three consecutive days. No bricks will be accepted which contain lime or other soluble substances in sufficient quantities to cause spalling or pitting of the surface when soaked in water for three consecutive days and then exposed to the air for a corresponding length of time.

"When the bricks shall have been delivered upon the roadway, samples may be selected at random therefrom for testing, which must meet the following requirements: The average crushing strength of two-inch cubes taken from any part of the brick shall not be less than 12,000 pounds per square inch.

"The modulus of rupture for transverse strength shall not be less than 1,600 pounds when calculated by the formula,

$$R = \frac{3\,W\,l}{2\,b\,d^2} \quad ,$$

R being modulus of rupture, W=load in pounds, b=breadth, d=depth, and l=length, all in inches.

"The specific gravity shall not be less than two and one-tenth when determined by the formula,

$$\text{Specific gravity} = \frac{W}{W' - W''}$$

Where W=weight of specimen freed from moisture before immersion, W'=weight of same after seventy-two hours' soaking, and W''=weight of same in water.

"In any test for determining the resistance to abrasion and impact, the loss of the brick shall not exceed two and one-fourth times that of Lithonia Georgia granite when subjected to like test.

"The material shall be subjected to inspection by the engineer placed in charge of the improvement, who will select samples, not exceeding five in number, for each of the tests required for determining absorption, crushing strength, transverse strength, and abrasion and impact, and cause the necessary specimens to be prepared, the tests to be made, and will accept or reject the material in accordance with the results of such trials. An allowance of twenty per cent may be made for variation of single specimens, but the average results shall be in accordance with the provisions herein set forth, or the material must be rejected, and removed at the contractor's expense. The tests may be repeated upon the arrival of different shipments, as frequently as may be necessary to insure the acceptance of only such material as shall comply with the provisions of this specification."

What Has Been Done.

Bidders were informed that proposals would be considered for any of the varieties of brick which had been tested, but that such as showed a crushing strength of less than 10,000 pounds per square inch, or a loss in the rattler test of more than three times that of the granite, would probably not be adopted. When the bids were opened it was found that proposals included Nos. 5, 6, 13, and 14 only, and the contracts were awarded for using No. 14. Subsequently it was decided to pave about 300 or 400 feet in length of the south end on the south contract with Nos. 5 and 6, and a corresponding length on the south end of the north contract with No. 13, which has been done.

The form given to the roadway is shown in the accompanying sections. Figure 1 showing the street where occupied by street railway tracks, and figure 2 the section where no tracks were placed. Figure 3 is a full-sized section of the rail adopted. This was placed directly on the tie, or rather on a bearing plate but three-eighths of an inch in thickness, spiked directly to the tie; each alternate plate having an outside brace to aid in holding the rails to guage, thus obviating the necessity of using any other appliance to effect that purpose. In surfacing tracks, the rail of each track nearest the center line of the street was placed one inch higher than the opposite rail of the same track. Templates were used for forming the sand bed on which the pavement was placed. Figure 4 shows that used for the the central space between tracks; figure 5 the one for the tracks, and figure 6 that for the spaces between car tracks and curb-stone. It will be noticed that the template used at the sides differs slightly from the form specified in the cross-section of street, but this variation is believed to be a betterment, being in the nature of a camber against the greater weight of traffic, and tending to confine the water near the curb.

Stakes were given for the alignment and grade of tracks,

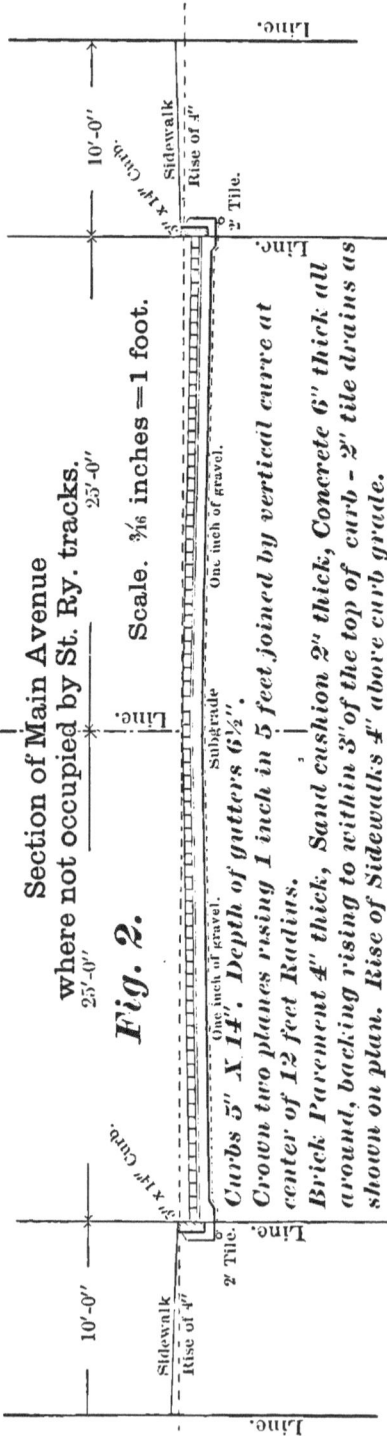

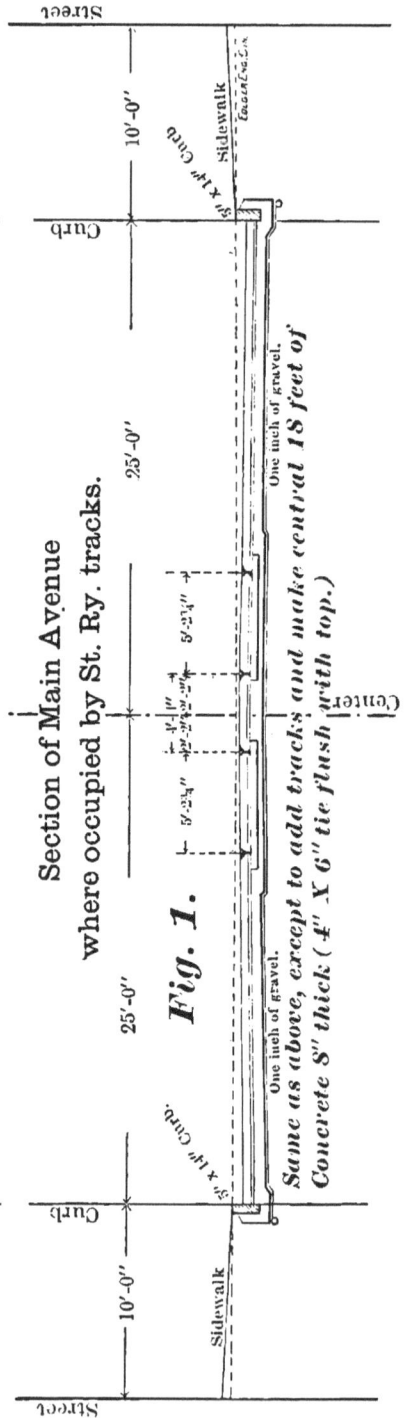

Fig. 2.

Section of Main Avenue
where not occupied by St. Ry. tracks.

Scale. 3⁄16 inches = 1 foot.

Curbs 5″ X 14″. Depth of gutters 6½″.
Crown two planes rising 1 inch in 5 feet joined by vertical curve at center of 12 feet Radius.
Brick Pavement 4″ thick, Sand cushion 2″ thick, Concrete 6″ thick all around, backing rising to within 3″ of the top of curb - 2″ tile drains as shown on plan. Rise of Sidewalks 4″ above curb grade.

Fig. 1.

Section of Main Avenue
where occupied by St. Ry. tracks.

Same as above, except to add tracks and make central 18 feet of Concrete 8″ thick (4″ X 6″ tie flush with top.)

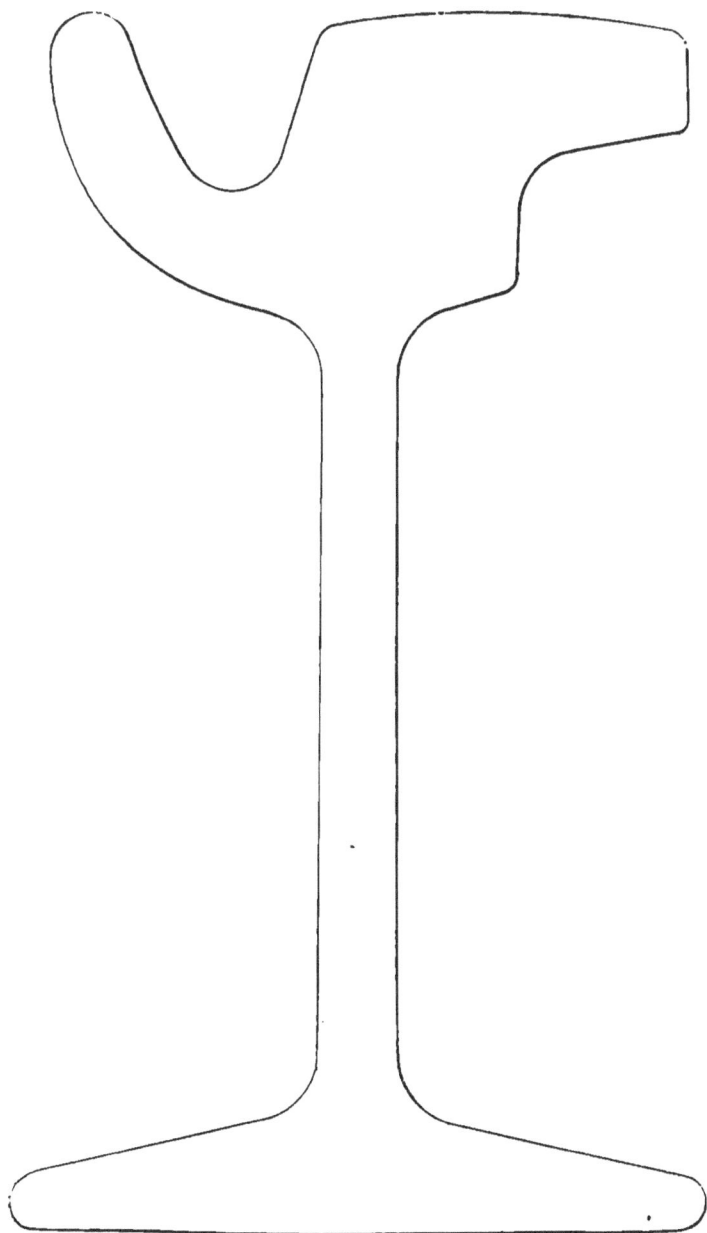

Fig. 3.

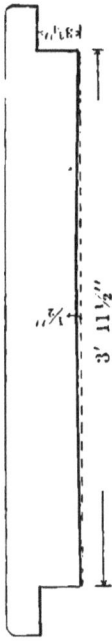

3' 11½"

Fig. 4

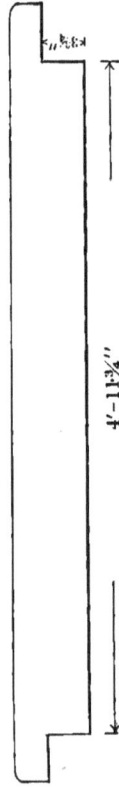

4'-1¾"

Fig. 5

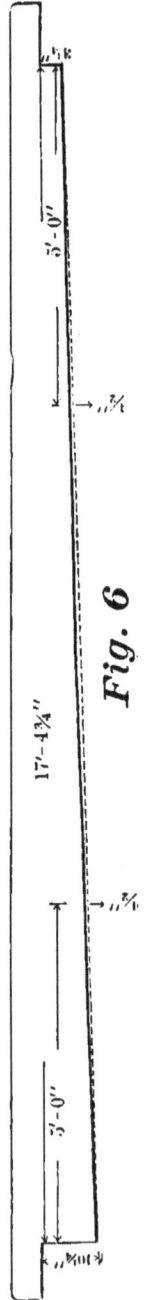

17'-4¾"

Fig. 6

and where no tracks were laid, levels were given from the curbs for forming the concrete and bed for the pavement. The resulting form of street has less crown than is usually given to streets paved with brick, but it is pleasing to the eye, the entire surface is available, and it is believed that it will be more durable than it would be were the cross grades steeper, because there is no influence tending to concentrate the traffic in lines on any portion of the roadway.

The form of rail-head will be observed to be almost identical with that recommended so urgently in my report of May, 1889, but the base of rail and general construction of track is materially different. The ties are of white oak, 5″ x 7″, placed three feet between centers, excepting at rail-joints, where the ties are double width. The joints are strong; splice bars being 24″ in length, fitting snugly against base and head of rails and being secured by six one-inch bolts. The pocket or space between base and head of rail was filled with a fine concrete or cement mortar, smoothed with a trowel to make vertical faces against which the paving was set. The concrete was driven under the ties with tamping bars, so as to secure a depth of three inches of this material on which they should rest. For the heavy motors and high speeds now used on electric roads, this depth should be increased, but when no part of the expense of such foundation is borne by the railway company, it appears unjust to increase the cost of a street improvement solely for that company's benefit. When grants to street railway companies are made for the use of such power, they should contain provisions for placing the tracks on suitable foundations, in order that stability may be obtained, and pavements not be endangered from such causes.

The form of curb adopted, and the manner of setting it, were very much criticised when the plans were first submitted, as it is a radical departure from what has been considered established city practice, but the innovation is now very generally conceded to be an improvement. Bottoms of curb stones were not required to be dressed or pitched to a line, and no spalling away required where variations from specified dimensions did not exceed two or three inches. There was a disposition to reject pieces varying from the prescribed dimensions when the depth at any point was less than the specified depth. To avoid having such pieces rejected, the quarrymen soon learned to confine the variance to excess only, which resulted in getting stones of an average depth of sixteen rather than fourteen inches. Inasmuch as they had been informed that pitching to a line would not be insisted upon where stones were broken to within two inches of prescribed dimensions, the result was a rather detrimental excess of stone. The tile drain was generally placed twenty-four inches below the top of curb and covered with sand, the top of which was smoothed off at just twenty inches below top of curb, then a foundation of concrete about four inches in depth was placed, upon which the curb-stones were blocked in position, and the residue of the concrete required by the plan tamped under and filled about them. This form of curb should reach to and be bedded in the concrete; excessive depth of stone merely serves to more nearly cut the concrete off and deprive the curb of efficient support. If exact dimensions are to be insisted upon, curbs should be cut to them on all sides, but that is obviously a useless expense.

There was almost constant irritation regarding the relative elevation of the pavement and the rails of the car tracks. The language of the specifications upon this point is the following: "The bricks or blocks will be paved adjacent to the rails in such manner that when rammed and rolled, as herein before specified, the surface of the pavement will not vary more than one-fourth ($\frac{1}{4}$) of an inch from that of the steel rails, and never be below them." For work to be done under similar circumstances, the last clause should read, "and never be *above* them," because, for some undefinable reason, this was construed to mean that the pavement was intended to be a quarter of an inch higher than the rail. Excessive ramming would drive the bricks down along the lines of rails, but there was no method of raising them except to take them up and put in sand. After being required to raise a few patches upon the mistaken idea that the bricks were too low, the pavers would raise the templates, keep the bed high, and endeavor to escape censure by ramming heavily along the rails, resulting in getting the surface of the pavement about one-fourth of an inch higher, as compared with the rails, than it should be. This will very greatly increase the wear upon the bricks adjacent to the tracks, enhance the tendency of the pavement to rise, or the bricks to "pop out" when shaken by passing cars, lessen the tractive force of the motors, add to the labor of keeping the rails clean, and lead vehicles to follow these incipient grooves along the car tracks, which the grooved rail was intended to discourage. No mention need be made in this connection of the causes which produced this result, since it can not now be obviated. Where the directions of an engineer carry with them the clear, undisturbed force which such orders should have, results are seldom variable, and if they are unsatisfactory the responsibility is not a divisible quantity. When the traffic tonnage is compared, it will be found that the eight inches of steel in the two car tracks carries so much more than an equivalent width of brick, that it is fair to assume that the rails will be lowered almost, if not entirely, as rapidly as the bricks will be worn down, so that there never is a time when a good reason can be given for

not making the surface approximate as closely as is practicable to a uniform plane. There is a much nearer approach
to this result on Main avenue, as it is, than is found in the
average city street, and, taken as a whole, it is believed to
be at this time the most comfortable as well as the hand-
somest drive which contains a double track street railway
line in the State of Ohio, and the street car service in the
village is believed to be as efficient, and approach as nearly
to the ideal rapid transit, as has yet been attained by any
suburb.

THE MATRIX.

One condition appears to be essential in the construction
of brick or stone block pavements, which are to be either
durable or smooth, and that is, to give each brick or block
such efficient support, and to bind all so firmly together,
that passing loads, whatever their weight, shall cause no move-
ment of the separate pieces upon the foundation or each other.
In brick pavements on a sand cushion this result can only be
secured by completely filling all joints with a cementing substance. On this work the substance used was coal-tar, or,
as it is generally called, paving cement. Care was taken to
keep the surface clean during the setting and ramming of
the bricks, to sweep all spalls or rubbish from the surface in
advance of the tar-pourers, and to project into each joint a
stream of tar at a high temperature that should penetrate
and fill every cavity from the sand-bed to the surface. A
second pouring or "plugging" always followed the first,
in order to fill such joints as allowed the tar to sink below
the surface to any perceptible extent, and then the surface
was covered with clean, sharp sand. Practically, the best py-
rometer for determining the temperature of the tar is its
appearance. Pouring should not be allowed unless it is
smoking hot and perfectly liquid. To assert that every one

of the millions of crevices was *completely* filled is not intended, since such a result would hardly be within the bounds
of possibility, but in every instance where bricks were taken
out after the tar-pouring had been done, they were found to
be solidly united by the tar, and hardly a square inch of
surface could be discovered that had not been covered by it.
Great care is necessary in doing the tar-pouring, as the
smearing of tar over the surface, leaving the spaces beneath
unfilled, is much worse than omitting it entirely, as its
presence will prevent the sand covering from penetrating
until the traffic shall have rendered the surface uneven. The
sand used for covering the pavement should be absolutely
free from loam. It should be sharp and gritty. There
should be a marked variation in size of the particles of
which it is composed, and they should be sharp and angular,
not rounded. One-fourth of an inch in depth is an abund-
ance in quantity, and it should be evenly spread as soon as
the tar-pouring has been completed, so that when the tar
sinks into the joints, if it does sink, it will carry the grit
with it.

A cement grouting is coming into general use in place
of the coal-tar filling, which is preferred by many engineers,
from the fact that hot weather renders the tar so nearly fluid
that it flows toward the lower levels, and leaves the higher
points unfilled. If made of Portland cement, and thoroughly
slushed into all crevices, the cement would give satisfactory
results. For a comparison between the two materials these
conditions exist. With the coal-tar, however much it may
be melted by heat, it reunites and becomes firm again so
soon as the temperature falls; while with the cement, should
the bond once be broken from any cause, it never reunites.

WHERE SHOULD BRICK BE USED FOR STREET PAVEMENTS.

Statements made in preceding paragraphs have indicated
that the brick pavement will not have the durability in localities where street traffic is greatly concentrated, that
would be obtained from the greater mass, and superior resistance to wear of granite blocks. Are bricks, then, a suitable paving material in any locality? Undoubtedly they
are, when they do not absorb more than two or three per
cent of their weight of moisture, and will withstand the wear
of the street traffic for a reasonable length of time. In any
city having a population less than 100,000, there are few, if
any, streets upon which the volume of traffic is concentrated
to such a degree as to endanger a well constructed brick
pavement for many years. Upon streets mainly occupied by
residences, where reasonable quiet and correct sanitary conditions are indispensable, brick are in every way superior to stone
block pavements, and in durability; when the cost of maintenance is considered are superior to asphalt, because the latter
form of pavement, when not used at all, will undergo changes
which will render it brittle and practically useless, whereas
the brick will not be similarly affected. This change in asphalt
from exposure is probably not thoroughly understood, but it
is acknowledged to exist. and is the principal cause of expense in maintaining pavements made of it. Sheet asphalt
is not worn from a pavement by the traffic to such an extent
that it would require renewal during the lifetime of a gen-
eration, but it is pressed out of position by the traffic, or
rendered brittle by changes in its composition. Probably
two-thirds or more of the cost of maintenance is due to the
latter cause, which action is believed to be retarded rather
than accelerated by the traffic upon it. In brick pavements
the action of the weather will affect the combining materials,

but not to an appreciable extent the bricks themselves unless
it be defective ones which can readily be removed and replaced. It follows, therefore, that the cost of maintenance
of brick streets, when made of good material on permanent
foundation and subjected to moderate traffic, will be less than
that of any other form of pavement, excepting only granite
blocks. Brick is, therefore, a suitable material to be used
upon a very large percentage of all paved streets, for granite
blocks are in no way fitted for use, either upon residence
streets or upon such avenues as lead to residence districts,
where the driving is at fairly high speed, and should be done
with safety and comfort. This brings brick into direct com-
petition with asphalt as a paving material, and, while it must
be admitted that the asphalt paving companies are deserving
of great credit for their efforts and success in producing good
streets, it must also be granted that persevering, honest com-
petition will result in public benefit if conducted on correct
principles. The indiscriminate careless use of brick as a
paving material will result beneficially to the asphalt interest, and disastrously to the brick manufacturers, but the construction of good brick streets will create an increasing demand for them. And while it may not seriously injure the
business of the asphalt companies, it will have a tendency
to lower both the first cost and the maintenance of streets
constructed by them.

Many streets are now opened and paved in a more or less
expensive manner, where no such outlay should be made, as
the work is totally wrecked in the subsequent laying of gas
and water mains, constructing sewers, and making connections with such lines. Common sense would indicate that
the order of work should be reversed; that a subgrade should
be made; all underground conduits completed and connections therewith carried to property lines; the surface then
covered with a coating of gravel or broken stone and gravel,

which should be used as a roadway until trenches had become firm, and this metal would then remain on which to place the foundation for the permanent pavement. Were this order of work insisted upon in opening new streets an immense saving would be effected. In repaving, the work should be done in the same order, suiting the manner of executing it to the necessities of the case, and the control of all of it should be under a single authority, with plans emitting from one source. When the traffic upon any street, or the use of such street demanded it, let such a pavement be laid as is suitable to meet the requirements of the case, and when it has been put down, keep it clean, in good repair and open to public use. Upon a very large percentage, probably more than two-thirds, of the paved streets in the larger cities, and upon nearly all streets requiring pavements in the smaller cities and towns, brick will be found to be a suitable material to compete with other substances for street paving. This is not saying that brick is the " coming pavement." On the contrary brick has got here. If used properly it will stay and will be a standard material for paving streets. In the great cities where traffic is heavily concentrated and little attention is given to the roaring of the passing current, it will never supplant the granite block; in the park drive, the clean rolled surface of hard finely broken stone will not give way to it; the fine avenue, where neatness, order and style must be maintained, will still be paved with asphalt, but it is to be hoped that the noisome rotting wooden block, and the rattling filthy cobble-stone, will not remain to annoy every sense of propriety and slowly murder both innocent and guilty by their noxious exhalations.

Upon nine-tenths of the paved streets occupied by residences, retail stores or office buildings. it would be an error almost criminal to put down the noisy, untidy granite block, and it would be still worse to use the wooden block, as it is

ordinarily laid in this country, while a *good* brick pavement
would be quiet, can be kept very clean, and in nearly all such
cases would last almost a lifetime, with a minimum expense
for repairs.

MAINTENANCE.

The ordinary practice regarding the maintenance of
street pavements appears to be at variance with established
methods in any other similar thing. When a line of railway
has been constructed it is placed in charge of the department
of maintenance of way, and is kept in proper condition for
use. When a street has been improved the gay and festive
plumber, singing as he toils, is legally *permitted* to dissect its
vitals; *licensed* sewer-tappers will disembowel it upon the
authority of a sovereign department; street railway companies will dig it up to adjust their tracks; the water department will probe it to rearrange their connections, and gas
companies will carve it to erect their lamps. Each will re-
place the disturbed material in his own way, and the street-
cleaning department will haul away such as remains loose
upon the surface. In the meantime no one looks for. or
remedies defects in their incipiency, the street having just
been paved is supposed to require no attention, and so long
as it remains passable without danger to life and limb, is not
repaired. If a drain becomes clogged no one knows any-
thing about it until the owners of the inundated properties
file claims for damages, which are promptly referred to the
engineer and solicitor. If vehicles are wrecked or animals
crippled the claims filed by owners go to the legal department, and not until the street has been absolutely destroyed
will it receive any attention from the repairing department.
If it should happen to have been constructed under the
supervision of officials of an opposite political complexion
from those now repairing it, money will be lavished upon it
to show how utterly rotten and useless were the works con-

6

structed at enormous expense by the other party. Should it
be some of their own work, it will be easy to show that it
was honestly constructed, but was ruined by the actions of
other sovereign and independent departments.

Imagine the effect of placing railways under the control
of half a dozen independent boards with no executive head,
their revenues separated into distinct funds with sovereign
boards to disburse them, each caring mainly that its minutes
shall record resolutions, ordinances, or references in proper
sequence and due form; so worded as to guard the rights
and actions of the board as a body, and show that various
matters were *considered*, and would be acted upon when some
other department had done something else. Think of any
corporation conducting any business enterprise upon such
methods, and cease to wonder why pavements are not kept
in repair.

During the first years of the life of a pavement it should
be carefully watched, and the beginnings of evil to it should
be checked, just as a new line of railway will require a
heavier force of section men than one that, having been
properly maintained, has been longer in use. Expensive re-
newals may be needed as portions of the structure become
worn by use, but care and watchfulness are of greatest use
upon new work. And this is just as true of streets as it is
of any like constructions. Contracts for street construction
frequently contain provisions requiring the constructor to
maintain his work for periods of time varying from one to
five or more years, but the meaning usually given this clause
by the contractor is that, at the expiration of the time
named he shall make such repairs of the portions of his
work which have not been dug up in the interval by some
other party as may be designated, and received the retained
percentage. There is uncertainty about this provision re-
maining in force for any considerable length of time. En-

terprising attorneys may argue that assessments should be
made for the cost of construction, and that the expense of
maintenance should not be assessed, but borne by the corporation, and no one can tell what the court will say until it
speaks in deciding the case as then presented. An efficient
force, under experienced, skillful direction, employed in the
inspection and *maintenance* of streets would appear to be an
absolute necessity in every municipality. If such an organi-
zation exists in *any* American municipality, it has published
no report of its services to date. If existing regulations can
not be bettered, then our form of government as applied to
municipalities is a failure.

WHAT IS IN A NAME.

The title *brick*, as applied to clay products used for street
paving, would appear to the writer as a misnomer. The
name ordinarily conveys to the engineer or builder the idea
of a brittle porous substance, so hungry for moisture that it
must be saturated before being laid in mortar, solely for the
preservation of the mortar, so brittle that unless combined
in masses it has little strength, and in no way suited to withstand the attrition or abrasion of street traffic. When people propose to use such a substance for paving streets, the
idea is ridiculed, and they must explain that they are not
using building brick, but an entirely different substance,
manufactured by brick makers, and in explaining the matter
use is made of the other unhappy term, " *vitrified brick.*"
The only clay product suitable for use in paving streets resembles a *tile* in more respects than it does a *brick*, and had
the name *tile* been chosen in the place of brick a more correct idea would have been conveyed. The first having gone
forth, however, it may be expedient to concur in the usual
practice, but it will always be necessary to bear in mind that
brick as used in street paving is a substance radically different from brick as used in any other connection.

SIZE OF PAVING BRICK.

A glance at the tables giving dimensions of specimens reveals the curious fact that hardly any two manufacturers make bricks of the same size. One of the first steps to be taken by manufacturers should be the adoption of a standard size for street paving bricks. Obviously their preference would be to make blocks of about the same dimensions as building bricks for both uses. When they make paving blocks only, and sell by the square yard, their interest will lie in the direction of increased thickness and diminished width. A large majority of manufacturers supply material for brick masonry as well as for paving, and can assort their output without material loss, thus enabling them to supply better goods for paving when they are required so to do, without suffering the entire loss of the value of such as may be rejected. This, from the manufacturers' standpoint, is the greatest argument in favor of making the dimensions of paving the same as building brick.

The *users* side of the question should be considered. The width of the brick or block forms the thickness or depth of the pavement. This should not be less than four inches. If made much in excess of that depth its cost will be increased about in the ratio of the increased depth. Inasmuch as four inches will afford ample strength and weight to resist the wear of the traffic to which this description of pavement is suited, there appears to be no reason for materially increasing the width beyond that named unless it be to meet exceptional cases. In the future, should it appear that brick are so perfected as to be able to carry the extremely heavy traffic concentrated upon business thoroughfares, where granite block pavements are now thought to be most suitable, a greater width may be found desirable. The length of the brick or block should be about twice its width;

its thickness should not exceed its width and may be made equal to it, providing such a block can be properly burned. The writer does not say that manufacturers *can not* properly dry and burn a brick three or four inches in thickness, but he does say that they *do* not do it. The conditions and the experience all indicate failure when massive pieces of clay are sought to be burned into bricks or blocks suitable for street paving. The nearest approach to success, has been attained by making the block hollow on the lower side in order to facilitate burning. For the solid block a thickness of two inches, or at most, two and one-half inches, is as great as should be attempted. Even where the clays can be melted or "vitrified" readily, there is great risk incurred in attempting to increase the thickness, for such clays usually contract greatly, and the outer surface is almost certain to be fixed or seared by the intense heat before the inner portions shall have been so acted upon as to produce the required vitrification. As a consequence they come from the kiln either insufficiently burned, checked with "fire cracks," either internal or external; or, like an ill-shaped casting, so affected by internal strains as to have no certain amount of strength. Better results are, therefore, likely to be secured by adopting about the building brick dimensions than by attempting to manufacture blocks of a larger size. Unfortunately those dimensions have never been determined with much accuracy in this country, but they should be, and then let manufacturers vary the dimensions of their molds as the contractility of the clays vary, so that bricks of equal hardness shall be of like dimensions.

It may be argued that the increased number of joints in a given area, caused by the thinner block, constitute an element of weakness and should, therefore, be avoided. The defect is more imaginary than real, since the proposition can not be true if made general. The perfect pavement would become one without joints, which is impracticable unless made of a substance sufficiently yielding or elastic to afford secure footing for animals, which practically makes it a surface of innumerable joints. An advantage claimed for brick pavements is said to be the fact that they can be so closely laid, and the joints so completely filled, that, while they furnish secure footing for animals, they are so smooth as to be quiet, and so impervious as to be cleanly. If this be true, the additional number of joints is not objectionable. They are not an element of weakness, since the load must in any case be carried by the foundation, and the upper and lower surfaces being equal, the weight transmitted by the brick will be as its area. Should the surface of the foundation be uneven, the smaller block is less liable to be tilted by an unequal pressure than the larger one. Within reasonable limits, therefore, the safe course to pursue would undoubtedly lie in the direction of the thinner blocks: or, in other words, to adopt a standard size for paving blocks corresponding with building brick dimensions.

www.ingramcontent.com/pod-product-compliance
Lightning Source LLC
Chambersburg PA
CBHW021422090426
42742CB00009B/1213